herapy
Kazushi Kishi

ed by InTech
dine 9, 51000 Rijeka, Croatia

hing Process Manager Vana Persen

ical Editor Teodora Smiljanic

Designer InTech Design Team

onal hard copies can be obtained from orders@intechopen.com

ytherapy, Edited by Kazushi Kishi

978-953-51-0602-9

BRACHY

Edited b

Brachy
Edited b

Publis
Janeza

Publi

Tech

Cove

Add

Brac
p. cr
ISBN

For EU product safety concerns:
IN TECH d.o.o., Prolaz Marije Krucifikse Kozulić 3, 51000 Rijeka, Croatia,
info@intechopen.com or visit our website at intechopen.com.

Meet the editor

Dr Kazushi Kishi holds the chair of department radiation oncology, tumor center, at the Wakayama Medical University Hospital, and is an Associate Professor of Department of Radiology at Wakayama Medical University, Wakayama City, Japan. His academic background includes PhD in radiology. He is a council of Japanese Society of Therapeutic Radiology (JASTRO), Japanese Radiological Society (JRS), Biological Group of JRS, and a member of American Society of Therapeutic Radiation Oncology and American Society of Brachytherapy. His current work is focused on interventional brachytherapy that expands the possibility of brachytherapy, i.e., to increase the therapeutic ratio of effect/toxicity, and enthusiastically developing new advantageous strategies, even to the difficult conditions as re-irradiation. He has published over more than 250 papers including 62 peer reviewed English journal articles. His work is currently supported by Grant-in-aid and now being widely recognized in the field of radiation oncology and cancer medicine.

Contents

Preface

Science of brachytherapy is currently increasing in the field of cancer therapy. By virtue of the development of computer assisted technologies, we can reap the benefits of accurate radiotherapies more than ever, which enabled precise irradiation to a target safely avoiding unnecessary involvement of the surrounding normal tissue. Unlike teleradiotherapy provided by X-ray linear accelerator, brachytherapy utilizes radiation source placing inside or close to the target. Taking advantage of practical availability of the physical characters including decay modes, half-lives, and decay energies of the radioisotopes, various practical insertion or deployment methods have been developed. Some of these are technically sophisticated. A precise treatment planning requires accurate dose calculation algorithm even in brachytherapy. In the clinical practice of brachytherapy, safe interventional procedures are indispensable. Reverse, a simultaneous intervention modifying the surrounding structure may be practically applicable.

In the first section, *In Physics of Brachytherapy*, Zehtabian et al. addresses the problem of currently popular dosimetric method (AAPM TG-43) based on experimental data. Orion et al. reports a potential benefit of Monte Carlo method in dose calculations of ruthenim-106/rhodium-106 (beta emitters) eyes applicators. In the second section, *Californium*, Wang present an excellent review of Progress in Californium-252 (Cf-252) Neutron Brachytherapy, including history, availability, fabrication, biological basis, dosimetry and clinical data including promising future possibilities. In the third section, *Practice of Brachytherapy*, Santini et al. reports their clinical case series study of high dose rate endobronchial brachytherapy in central airway tumor obstruction. Impressive result indicate a singinficant rationale of brachytherapy in this situation. Zeituni et al. elaborate a comprehensive review from principle and techniques of Iodine-125 production at nuclear reactor plant to their clinical practive in prostate cancer treatment. In the last section, *Novel Strategy and Applications*, Ruge et al present a very interesting review with their clinical practices of Stereotactic Brachytherapy for Brain Tumors using Iodine-125 seed. We may reappraise our current concepts in the treatment of gliomas that has never been satisfactory controlled. In the last chapter, Kishi et al introduce a brachytherapy procedure with organ-sparing hyaluronate gel injection as a simultaneous intervention, which enabled safe and eradicative reirradiation, by modifying the surrounding structure during the irradiation.

Though this book is not an unabridged review of brachytherapy, it is full of essences to make a breakthrough in radiation oncology by brachytherapy. I hope this book will encourage all people from related fields.

Acknowledgement

I would like to thank all contributors, all people related and especially Ms. Vana Persen, the wonderful Publishing Process Manager, my mentors, and my family who have supported me throughout the term.

Dr Kazushi Kishi
M.D., Ph.D., Department of Radiology,
Wakayama Medical University,
Japan

Section 1

In Physics of Brachytherapy

1

A Review on Main Defects of TG-43

Mehdi Zehtabian[1], Reza Faghihi[1,2] and Sedigheh Sina[1,2]
[1]*Nuclear Engineering Department, School of Mechanical Engineering, Shiraz University,*
[2]*Radiation Research Center, School of Mechanical Engineering, Shiraz University,*
Iran

1. Introduction

The Sivert integral and modular dose calculation models (TG-43) are two methods used for calculation of dose distributions around a brachytherapy source (Khan, 2003). Among all methods used for calculation of dose distributions, TG-43 has become most popular and most promising method because it uses the quantities measured in the medium to calculate dose rates.

In this chapter, we will explain the TG-43 dose calculation formalism and after verification of the main reason of its popularity, we will review main defects of this formalism.

1.1 TG-43 formalism

In 1995, the American Association of Physicists in Medicine (AAPM)) published a report on the dosimetry of sources used in interstitial brachytherapy, Task Group No. 43 (TG-43) (Nath et al., 1995). This report introduces dose calculation formalism utilizing new quantities like air kerma strength (S_K), dose rate constant (Λ), geometry function (G (r,θ)), radial dose function (g (r)), and anisotropy function (F (r,θ)). These dosimetry parameters consider geometry, encapsulation and self-filtration of the source, the spatial distribution of radioactivity within the source, and scattering in water surrounding the source. According to this protocol, the absorbed dose rate distribution around a sealed brachytherapy source at point P with polar coordinates (r, θ) can be determined using the following formalism:

$$D^{\cdot}(r,\theta) = \Lambda\, S_k \frac{\mathrm{G}\,(\mathrm{r},\theta)}{\mathrm{G}\,(\mathrm{r}_0,\theta_0)}\, g(r) F\,(r,\theta) \qquad (1)$$

where r is the distance to the point P and θ is the angle with respect to the long axis of the source, and (r_0, θ_0) is reference point that r_0= 1 cm and $\theta_0 = \pi/2$ (Fig. 1).

1.1.1 Air kerma strength

Air kerma strength, S_k, acounts for brachytherapy source strength and is defined as the product of air kerma rate at a calibration distance (d) in free space, which is usually chosen to be 1 m, along the transverse axis of the source and the square of the distance (d^2),

$$S_k = K\,(d) d^2 \qquad (2)$$

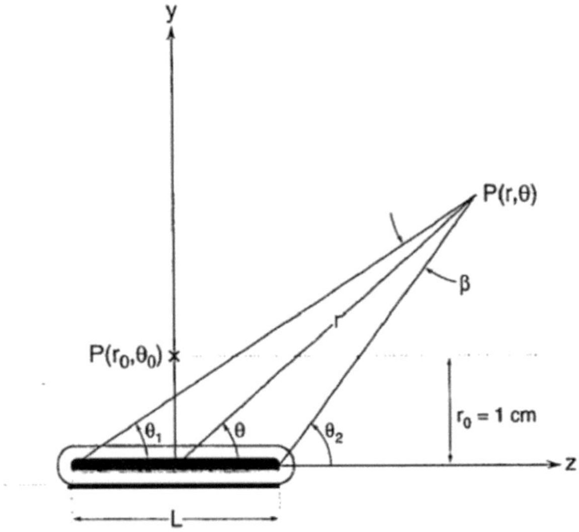

Fig. 1. The geometry of brachytherapy sources used in TG-43 formalism (Nath et al., 1995).

Where K (d) is the air kerma rate, and d is the calibration distance (d). The unit of S_K is U (1 U =1 cGy cm^2 h^{-1}).

1.1.2 Dose rate constant

Dose rate constant, Λ, is defined as dose rate at 1 cm along transverse axis ($\theta_0=\pi/2$) of the source per unit air kerma strength (U) in a water phantom. Dose rate constant of a brachytherapy source is obtained as below:

$$\Lambda = \dot{D}\ (r_0, \theta_0)/S_k \tag{3}$$

Λ has units of cGy h^{-1} U^{-1}. As indicated in Task Group No. 43, the effects of source geometry, the spatial distribution of radioactivity within the source, encapsulation, and self-filtration within the source and scattering in water surrounding the source is considered by this parameter.

1.1.3 Geometry function

Spatial distribution of radioactivity within the source and the slump of the photon fluence with distance from the source are considered by Geometry function, G (r, θ), with the unit of cm^{-2}.

For point sources, the geometry function indicates the inverse square law.

$$G(r,\theta) = \begin{cases} r^{-2}, Point\ source \\ \dfrac{\beta}{Lr\ sin\theta}, Line\ source \end{cases} \tag{4}$$

As it is shown in Figure 1, L is the source active length and β =θ_2-θ_1 is the angle covering active source from the point (r, θ).

1.1.4 Radial dose function

Radial dose function, g (r), takes into account slump of dose rate arising from absorption and scattering in the medium on the transverse axis of the source and can be affected by self filtration, and encapsulation.

According to AAPM Task Group 43, radial dose function is defined as:

$$g(r) = \frac{\dot{D}(r,\theta_0)G(r_0,\theta_0)}{\dot{D}(r_0,\theta_0)G(r,\theta_0)} \quad (5)$$

This quantity is defined just on transverse axis, i.e., $\theta_0=\pi/2$. As it is obvious from equation 9.5, effect of the geometric falloff of the photon fluence with distance from the source (inverse square law) has been suppressed by G (r_0, θ_0)/G (r, θ_0).

1.1.5 Anisotropy function

Angular variation of photon absorption and scattering in the encapsulation and the medium at different distances and angles from the source is taken into account by Anisotropy function, F (r, θ).

$$F(r,\theta) = \frac{\dot{D}(r,\theta)G(r,\theta_0)}{\dot{D}(r,\theta_0)G(r,\theta)} \quad (6)$$

Where $\theta_0=\pi/2$.

Like g (r), applying geometry function in above equation is suppressing the effect of inverse square law on the dose distribution around the source.

Since the publication of TG-43 formalism, many investigations have been performed on determination of the dosimetry parameters of brachytherapy sources and verification of its advantages and limitations of this protocol (Liu et al.,2004; Meigooni et al., 2003, 2005; Melhus & Rivard, 2006; Parsai et al., 2009; Rivard et al., 2004, 2007; Sina et al., 2007, 2009, 2011; Song & Wu, 2008; Zehtabian et al., 2010).

1.1.6 TG-43U1

An updated version of TG-43 (named TG-43U1) protocol were published on 2004, adding several corrections to the original protocol (Rivard et al., 2004).

$$\dot{D}(r,\theta) = \Lambda\, S_k \frac{G_L\,(r,\theta)}{G_L\,(r_0,\theta_0)}\, g_L(r) F\,(r,\theta) \quad (7)$$

Equation (9-10) includes additional notation compared with the corresponding equation in the original TG-43 formalism, namely the subscript "*L*" has been added to denote the line source approximation used for the geometry function.

The definition of S_k in TG-43U1 differs in two important ways from the original AAPM definition of Sk.

The quantity Air-kerma strength, S_K , is the air-kerma rate, K_δ (*d*), *in vacuo* and due to photons of energy greater than δ, at distance *d* from , multiplied by the square of the distance which should be located on the transverse plane of the source. This distance (*d*) can be any distance large relative to the maximum linear dimension of the radioactivity distribution typically of the order of 1 meter.

$$S_k = K_\delta (d) d^2 \tag{8}$$

When S_k is obtained experimentally, the measurements should be corrected for photon attenuation and scattering in air and any other medium interposed between the source and detector, as well as photon scattering from any nearby objects including walls, floors, and ceilings (Rivard et al., 2004).

The low-energy or contaminant photons (e.g., characteristic x-rays originating in the outer layers of steel or titanium source cladding) would increase K_δ (*d*) without contributing significantly to dose at distances greater than 0.1 cm in tissue. Therefore a cutoff energy δ (i.e. 5 keV for low-energy photon emitting brachytherapy sources) should be considered in calculating K_δ (*d*).

In TG-43U1 dose-calculation formalism, a subscript X has been added to the notation g (r) and geometry function of the original protocol. This protocol presents tables of both g_P (*r*) (point source approximation) and g_L (*r*) (Line source approximation) values. g_X (*r*) is equal to unity at r_0=1 cm.

$$g_X(r) = \frac{\dot{D}(r,\theta_0)G_X(r_0,\theta_0)}{\dot{D}(r_0,\theta_0)G_X(r,\theta_0)} \tag{9}$$

The 2D anisotropy function, *F* (*r*,θ), is defined as:

$$F(r,\theta) = \frac{\dot{D}(r,\theta)G_L(r,\theta_0)}{\dot{D}(r,\theta_0)G_L(r,\theta)} \tag{10}$$

The definition of 2D anisotropy function is identical to the original TG-43 definition, other than inclusion of a subscript *L*, which is added to geometry function.

Generally the TG-43U1 includes:

a. A revised air-kerma strength (S_K) definition.
b. Not using *apparent activity* for specification of source strength
c. Distance-dependent one-dimensional anisotropy function
d. Guidance on extrapolating tabulated TG-43 parameters to long and short distances
e. Minor correction of the original protocol and its implementation (Rivard et al., 2004).

2. Limitations of TG-43 and TG-43U1 dose calculation formalisms

The dosimetry parameters used in the AAPM TG-43 and TG-43U1 dosimetry formalisms are obtained for a single brachytherapy source located at the center of a fixed volume, homogeneous liquid water phantom (Nath et al., 1995; Rivard et al., 2004, 2007). Consequently, these formalisms does not readily account for several aspects that undermine acquisition of high-quality clinical results (Rivard et al., 2009). Some important limitations of TG-43 and TG-43U1 are listed in the following sections (section 9.3.1 and 9.3.4).

2.1 The dosimetry phantom

As mentioned previously, the TG-43 parameters of a brachytherapy source are obtained in a homogeneous water phantom, but in actual clinical cases, the brachytherapy sources are located inside the tissues of the patients.

The different mass absorption coefficients, radiation scattering and attenuations in materials with different compositions would alter the dose distribution in comparison with water.

For example, the brachytherapy sources are placed inside the soft tissue, which is almost equivalent to water, with some small differences in density, atomic number Z_{eff} and chemical composition. Such differences between the compositions of soft tissue and water may cause some discrepancies between the TG-43 parameters in phantoms made up of such materials.

There are also other tissues inside the human body with more differences in density, atomic number and chemical compositions (i.e. bone, breast, lung, ...), for which much more discrepancies are observed in TG-43 parameters compared with water phantom. Table 1, shows the chemical compositions and densities of some tissues according to ICRU 44 (ICRU, 1989).

		bone	4 component soft tissue	Muscle
Density (g/cm3)		0.92	1.04	1.04
Element	Atomic number			
(H_2)	1	6.3984%	10.1172%	10.1997%
(C)	6	27.8%	11.1%	12.3%
(N_2)	7	2.7%	2.6%	3.5%
(O_2)	8	41.0016%	76.1828%	72.9003%
(F)	9			
(Na)	11			0.08%
(Mg)	12	0.2%		0.02%
(Si)	14			
(P)	15	7%		0.2%
(S)	16	0.2%		0.5%
(Cl)	17			
(K)	19			0.3%
(Ca)	20	14.7%		

Table 1. The chemical compositions and densities of some tissues (ICRU, 1989)

The effect of phantom material on dose distribution around brachytherapy sources and TG-43 parameters increases with decreasing the photon energy because of the dependence of photoelectric effect on photon energy and the effective atomic number of the absorbing materials.

The ratios of photon mass absorption coefficient of different tissues (μ_{en} tissue) to the mass absorption coefficient (μ_{en} water) of water ($\left(\frac{\mu_{en}}{\rho}\right)_{water}^{tissue}$) are shown on Table 2 for some energies (20, 30, 100, 300, 600, and 1000 keV).

Energy (keV)	Bone	fat	Muscle
20	4.57	0.56	1.03
30	4.80	0.58	1.04
100	1.12	0.96	1.00
300	0.97	1.00	0.99
600	0.96	1.00	0.99
1000	0.96	1.00	0.99

Table 2. The ratio of mass absorption coefficients of different tissues to water ($\left(\frac{\mu_{en}}{\rho}\right)_{water}^{tissue}$) (Khan, 2003).

As it can be seen from Table 2, the difference between mass absorption coefficients in phantoms are more pronounced sources of low energy photons.

To compare the dose distribution in diffeternt dosimetry phantoms, we performed MCNP4c (UT-Battelle & LLC, 2000) Monte Carlo calculations of The Selectron low dose rate (LDR) Cs-137 pellet sources of Nucletron (Nucletron BV, Netherland) (Nucletron, 1998) in homogeneous cubical water phantom of dimention 30*30*30 cm to obtain the dose distribution around a combination of 10 active pellets inside the cylindrical applicator, and then we repeated the MC simulations for different phantom materials (i.e. bone, muscle and soft tissue). The small cubical lattice of 1 mm dimension was defined in the phantoms in order to score dose distribution around the sources inside homogeneouse phantoms using tally F6.

Figure 2, compares the dose distribution around ten active pellets in a cubical water phantom with the dose distribution in bone phantom. The active spherical sources and inactive pellets along with the components of the cylindrical applicator can be seen on the figure.

According to the results of MC simulations, the percentage difference between the dose in water and bone phantoms at transverse plane of the source, increase by increasing the distance from the source center. For instance the percentage difference between the dose in bone and water at a point located at r=0.5 cm on transverse axis of the source is 3%, but this value increases to 20% at r=10 cm.

The comparison between the dose distribution around Cs-137 pellets in water phantom and other phantoms (soft tissue, and muscle) shows the percentage difference of about 3% at r=10 cm which are less pronounced than the percentage difference between bone and water, this is because of the difference in density and effective atomic numbers of these tissues (see Table 1) which lead to different mass attenuation and absorption coefficient of photons.

Melhus et al 2006 investigated the ratio of g (r) in different phantoms (i.e. ICRU 44, four-component soft tissue, breast, Muscle, and soft tissue) to g (r) in water phantom for different brachytherapy sources (i.e. I-125, Cs-137, Ir-192, Pd-103, and Yb-169) using MCNP5 Monte Carlo code (Melhus & Rivard, 2006).

The ratios of $g\ (r)_{tissue}$ to $g\ (r)_{water}$ are shown for different brachytherapy sources are shown in Figure 3 (Melhus & Rivard, 2006).

According to Figure 3, the difference between $g\ (r)_{tissue}$ and $g\ (r)_{water}$ increase with decreasing photon energy.

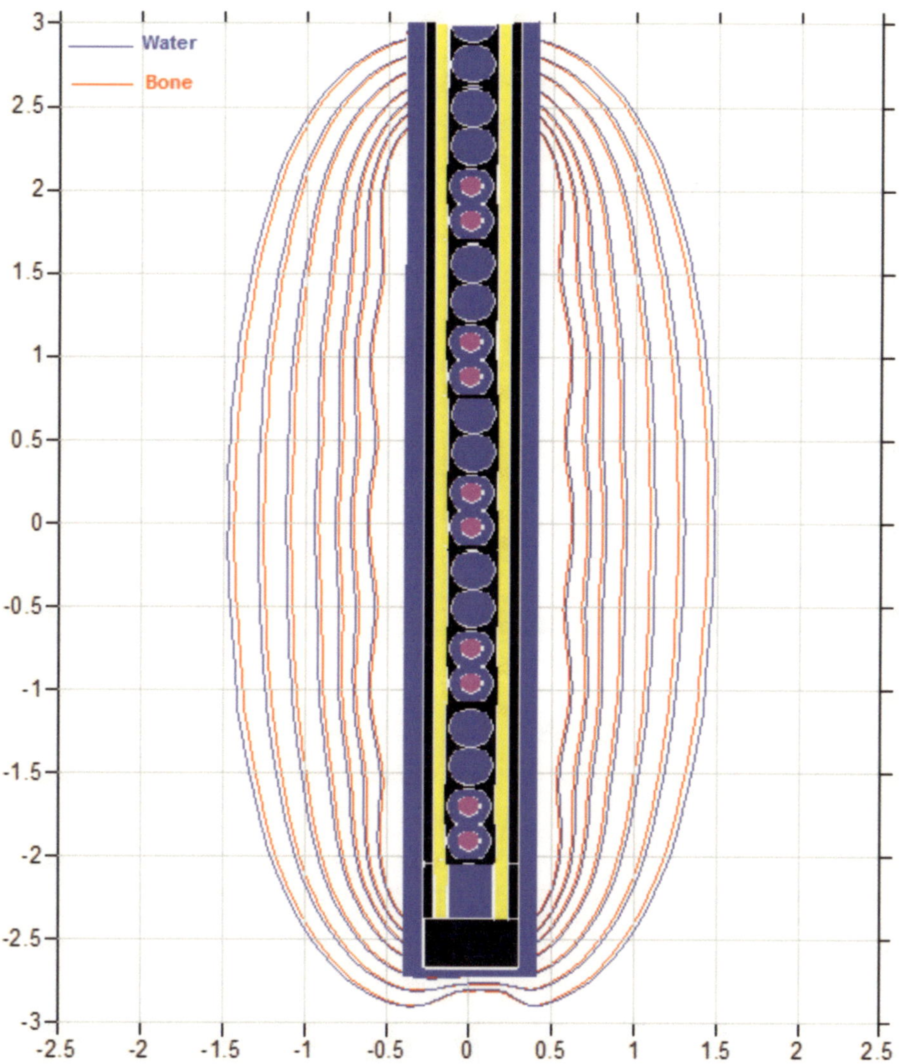

Fig. 2. Comparison of the dose distribution around a combination of 10 Cs-137 active pellets in homogeneous water phantom and bone phantom.

Melhus et al 2006, reported a factor of 2 differences in g (r) between water and ICRU 44 breast tissue for ^{103}Pd at a depth of 9 cm, and the difference of approximately 30% between g (r) for ICRU muscle and soft tissue at a depth of 9 cm for both ^{125}I and ^{103}Pd.

2.2 Inhomogenities

According to the TG-43 dose calculation formalism, the dosimetry is performed in a uniform medium (water) phantom and inhomogenities like bony and soft tissues were not taken into account. There are some implant sites such as head and neck and lung in which the existing inhomogenities are important and would change the dosimetry parameters of the brachytherapy source.

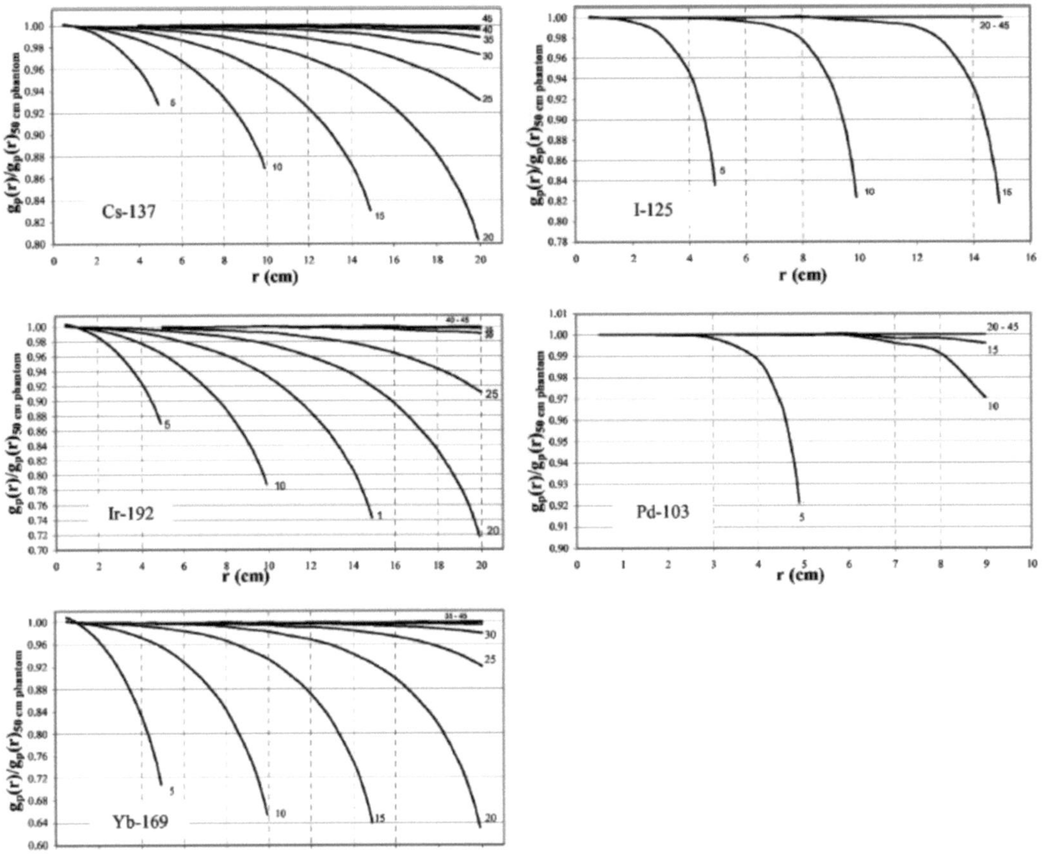

Fig. 3. The ratio of g (r) in different tissues to g (r) in water, for different brachytherapy sources, (i.e. I-125, Cs-137, Ir-192, Pd-103, and Yb-169) (with permission of Med Phys Journal) (Melhus & Rivard, 2006).

In 2008, Song et al performed a study on AAPM TG-43 dose calculation formalism in heterogeneous media for three radioactive seeds (^{125}I, ^{192}Ir and ^{103}Pd). They recommended the modified dose rate constants in different heterogeneities (**Λ**' (bone)=5.030 and 4.06 cGy/hU for I-125 and Pd-103 respectively) (Song & Wu, 2008).

In 2010 we performed investigation on the effects of tissue inhomogeneities on the dosimetric parameters of a Cs-137 brachytherapy source using the MCNP4C code (Zehtabian et al., 2010). We defined an updated dose rate constant parameter, for Cs-137 source in presence of tissue inhomogenities. The MCNP4C simulations were performed in phantoms composed of water, water-bone and water-air combinations (see Figure 4). The values of dose at different distances from the source in both homogeneous and inhomogeneous phantoms were estimated in spherical tally cells of 0.5 mm radius using the F6 tally. The percentages of dose reductions in presence of air and bone inhomogenities for the Cs-137 source were found to be 4% and 10%, respectively. Therefore, the updated dose rate constant (Λ) will also decrease by the same percentages. The dose rate constant of a single active pellet inside the liquid water phantom was 1.093. The modified values of dose rate constant (Λ'_a) for air and bone inhomogenitis were found to be 0.984 and 1.047

respectively (Zehtabian et al., 2010). By comparing our results and Song et al results, it can be easily concluded that such dose variations are more noticeable when using lower energy sources such as Pd-103 or I-125 because of prevail of photoelectric effect.

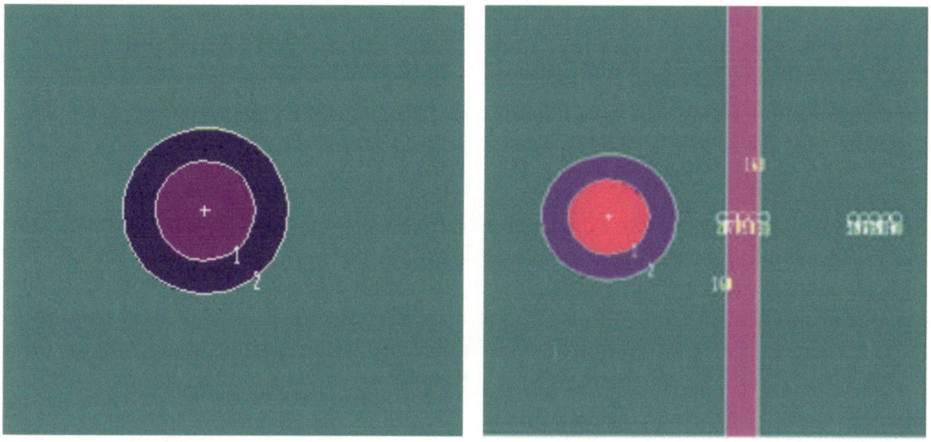

Fig. 4. Simulated geometry of a single active Cs-137 pellet inside the phantom a) uniform water phantom b) Water phantom with inhomogenity (bone or air) (with permission from Iranian Journal of Medical Physics) (Zehtabian et al., 2010).

2.3 InterSource Attenuation (ISA) and influence of applicators

For clinical applications, in which several brachytherapy sources are used in treatment of the patients, the ISA may be significant.

TG-43 parameters of brachytherapy sources are obtained for single brachytherapy sources inside a homogenious water phantom, not considering the effect of other brachytherapy sources (ISA) and the applicators (Markman et al., 2001).

The structures of brachytherapy sources and the applicators are composed of materials with the densities and atomic numbers different from that of water. The photoelectric effect is more dominant in materials with higher atomic numbers, so the TG-43 parameters are different when shielding effects of the applicator and other source are considered in comparison to the water without consideration of these factors (Rivard et al., 2009).

The radio-opaque markers (i.e. silver markers Z=47) which are used in identification during post Implant CT localization in LDR brachytherapy may also cause some inaccuracies in TG-43 dose calculation formalism.

The applicators which are made up of materials with the atomic numbers different from that of water (i.e. stainless steel=26) would also cause some errors in TG-43 parameters.

The difference between the TG-43 parameters of brachytherapy sources in applicator materials are highly dependent on the energy of brachytherapy sources. Consequently not considering the ISA and the applicator effects are more pronounced for brachytherapy sources emitting low energy gamma rays.

Different investigators have studied the shielding effects of other materials (such as the other sources, the applicator components and radio opaque markers) and showed that not taking the radiation attenuation in such materials into consideration may have significant effect on dose distribution around the brachytherapy sources (Margarida Fragoso, 2004; M. Fragoso et al., 2004; Parsai et al., 2009; Perez-Calatayud et al.,2004, 2005; Sina et al., 2007, 2011; Siwek et al., 1991).

To investigate the shielding effect of the cylindrical vaginal applicator and dummy pellets on TG-43 dosimetry parameters of Cs-137 low dose rate brachytherapy source, we simulated a single active pellet inside the applicator and calculated the geometry parameters of the source (i.e. dose rate constant, radial dose function and anisotropy function) (Sina et al., 2009, 2011a, 2011b).

According to the results of our study, the presence of the applicator does not have any significant effect on the dose rate constant and radial dose function of the source (θ=90°), but the value of anisotropy function differs from unity especially towards the longitudinal axis of the source (θ=0° or θ=180°). Figure 5 shows the anisotropy function of an active pellet inside the applicator (while other pellets are inactive) at different distances from the center of the source. As it is obvious from Figure 5, the value of F (r,θ) are equal to unity at θ=90°, and decrease as the angles reaches 0° and 180°.

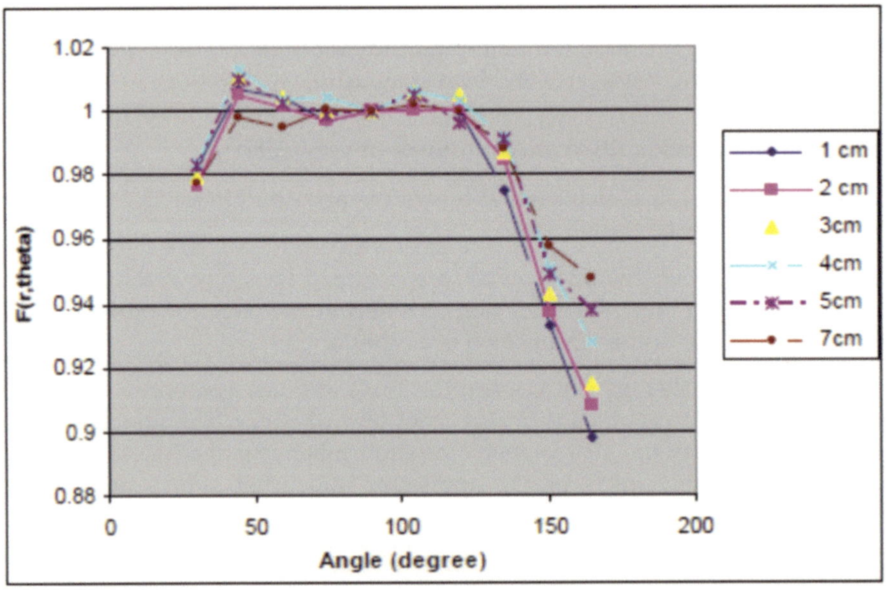

Fig. 5. Comparison of the anisotropy function at different distances and angles from the source (with permission of IJRR). (Sina et al., 2009)

2.4 Phantom size

Differences between radiation scattering for dataset acquisition and patient treatment would also cause some errors in dose calculations based on TG-43 formalism. In actual clinical cases the brachytherapy sources is not placed exactly at the center of the patient's body. Sometimes the radioactive source is placed near the patient's contour, and this change in the scattering condition inside the phantom, would have significant effects on TG-43 parameters of the sources.

The percentages of difference in TG-43 parameters are highly affected by energy of the photons emitted from brachytherapy sources, and the amount of missing tissue (the phantom size).

Several investigators has studied the phantom size as an important point in BT dosimetry study is the phantom size involved either in MC calculations or experimental measurements (Anagnostopoulos et al., 1991, 2003; Herbold et al., 1988; Karaiskos et al., 2003; Sakelliou et al., 1992; Tiourina, 1995). Williamson 1991, showed 5%–10% differences at distances greater than 5cm from a ^{192}Ir source comparing the dose in an unbounded liquid-water and a phantom of of nearly 20*20*20 cm^3 (Williamson, 1991). Venselaar *et al.* have investigated the influence of phantom size on dose by changing the water level in a cubic water tank for ^{137}Cs, ^{60}Co and ^{192}Ir sources (Venselaar et al., 1996). They obtained significant differences in the dose between experiments with different phantom sizes. Karaiskos *et al.* investigated the effect of water spherical phantom size on ^{192}Ir MicroSelectron HDR source of diameters ranging from 10 to 50 cm on the radial dose function, g (r), at radial distances near phantom boundaries, both experimentally and theoretically, and observed up to 25% differences at the points near the phantom boundaries (Karaiskos et al., 2003). It is better to take such differences into account in the dose calculations of the treatment planning softwares in clinical cases which the sources are located near the phantom boundaries.

Pe´rez-Calatayud et al verified the influence of phantom size on the radial dose function of different brachytherapy sources using GEANT4 code. They found that a 15 cm radius water phantom ensures full scatter conditions up to 10 cm from the low energy photon emitting brachytherapy sources like ^{103}Pd, and ^{125}I (Perez-Calatayud, 2004). They also suggested the unbounded phantom radius of 40cm for ^{137}Cs and ^{192}Ir sources.

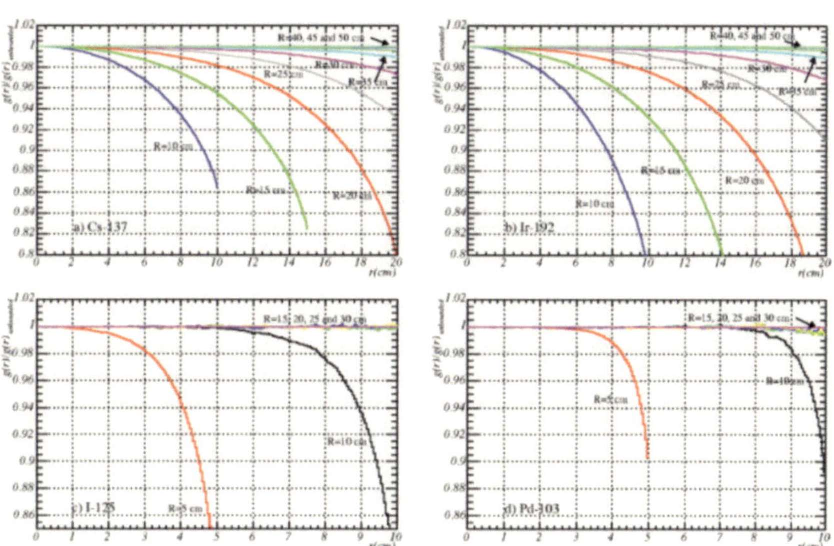

Fig. 6. Ratio between the radial dose function for a phantom size R and the radial dose function for an unbounded phantom, for ^{137}Cs, ^{192}Ir, ^{125}I and ^{103}Pd point sources and for various phantom sizes (Perez-Calatayud, 2004).

Melhus et al 2006, also compared the g (r) values for different phantom sizes with the values of g (r) for a phantom size of R=50cm using MCNP5 Monte Carlo code and found that the MCNP5 results were in agreement with the GEANT4 code (see Figure 7).

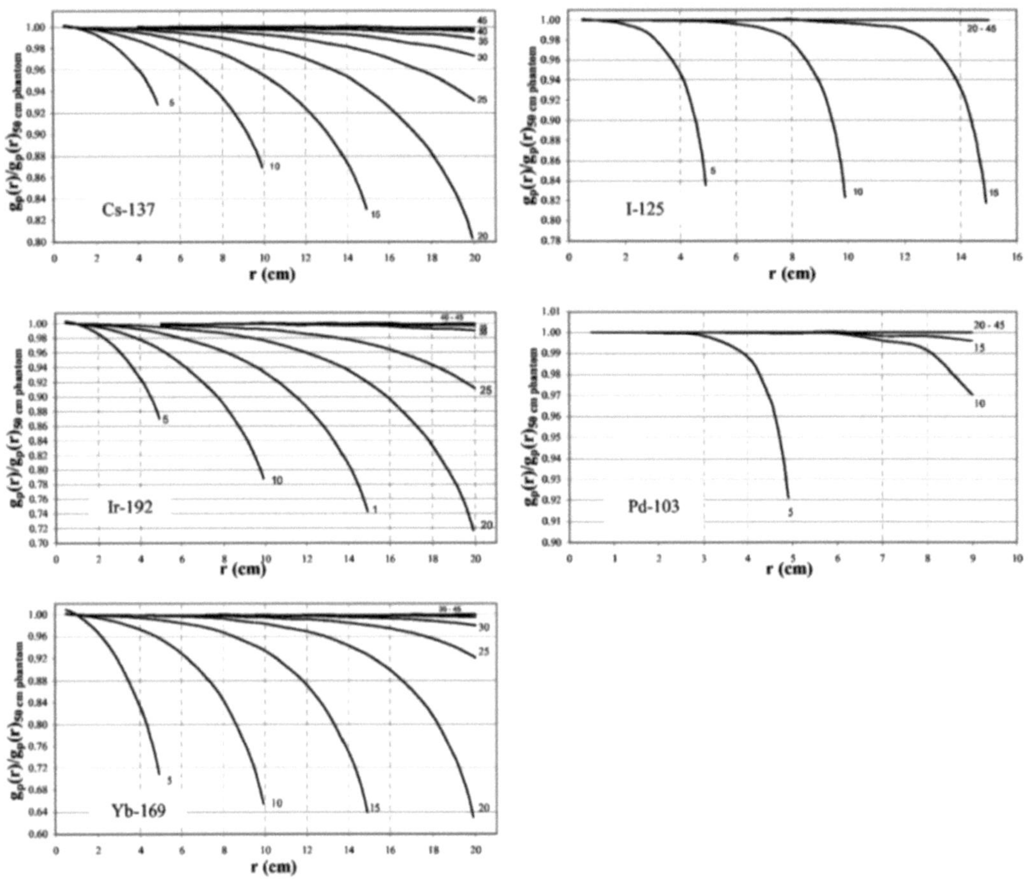

Fig. 7. Comparison of g (r) values for phantoms of different sizes with the phantom of R=50cm, for different brachytherapy sources. (Melhus & Rivard, 2006)

3. Conclusions

Today, most of brachyherapy treatment planning systems are based on the recommendations of AAPM Task Group #43 dose calculation formalism (TG-43). The tabulated data including TG-43 parameters of the brachytherapy sources which are obtained experimentally or theoretically, are used as the input data of the treatment planning softwares. The TG-43 parameters are obtained by positioning the source at the center of a fixed volume homogeneous water phantom, not considering the different scattering and attenuation of the photons in other tissues, the phantom size and the ISA and the applicator shielding effects.

Such limitations would affect the outcome of the treatment planning softwares. It is recommended that such factors be taken into considerations as much as possible, by applying some correction factors.

4. References

Anagnostopoulos, G.; Baltas, D.; Karaiskos, P.; Pantelis, E.; Papagiannis, P. & Sakelliou, L. (2003). An analytical dosimetry model as a step towards accounting for inhomogeneities and bounded geometries in 192Ir brachytherapy treatment planning. *Phys. Med. Biol.* , 48, 1625-1647.

Angelopoulos, A.; Perris, A.; Sakellariou, K.; Sakelliou, L.; Sarigiannis, K. & Zarris, G. (1991). Accurate Monte Carlo calculations of the combined attenuation and build-up factors, for energies (20–1500 keV) and distances (0–10 cm) relevant in brachytherapy,. *Phys. Med. Biol.* , 36, 763–778.

Fragoso, M. (2004). Application of Monte Carlo Techniques for the Calculation of Accurate Brachytherapy Dose Distributions., University of London, London,UK.

Fragoso, M.; Love, P.A.; Verhaegen, F.; Nalder, C.; Bidmead, A.M.; Leach, M., et al. (2004). The dose distribution of low dose rate Cs-137 in intracavitary brachytherapy: comparison of Monte Carlo simulation, treatment planning calculation and polymer gel measurement. *Phys Med Biol*, 49 (24), 5459-5474.

Herbold, G.; Hartmann, G.; Treuer, H. & Lorenz, W.J. (1988). Monte Carlo calculation of energy build-up factors in the range from 15 keV to 100 keV, with special reference to the dosimetry of 125I seeds. *Phys. Med. Biol.* , 33, 1037-1053. ICRU, I.C.o.R.U.a.M.R.N. (1989). Tissue substitutes in radiation dosimetry and measurement.

Karaiskos, P.; Angelopoulos, A.; Pantelis, E.; Papagiannis, P.; Sakelliou, L.; Kouwenhoven, E., et al. (2003). Monte Carlo dosimetry of a new 192Ir pulsed dose rate brachytherapy source. *Med Phys*, 26 (4), 9-16.

Khan, F.M. (2003). The Physics of Radiation Therapy (Third ed.). Philadelphia: Lippimcott Williams and Wilkins.

Liu, L., Prasad, S.C., & Bassano, D.A. (2004). Determination of 137Cs dosimetry parameters according to the AAPM TG-43 formalism. *Med Phys*, 31 (3), 477-483.

Markman, J.; Williamson, J.F.; Dempsey, J.F. & Low, D.A. (2001). On the validity of the superposition principle in dose calculations for intracavitary implants with shielded vaginal colpostats. *Med Phys*, 28 (2), 147-155.

Meigooni, A.S.; Rachabatthula, V.; Awan, S.B. & Koona, R.A. (2005). Comment on "Update of AAPM Task Group no. 43 report: A revised AAPM protocol for brachytherapy dose calculations". *Med Phys*, 32 (6), 1820-1821; author reply 1822-1824.

Meigooni, A.S.; Zhang, H.; Perry, C.; Dini, S.A. & Koona, R.A. (2003). Theoretical and experimental determination of dosimetric characteristics for brachyseed Pd-103, model Pd-1, source. *Appl Radiat Isot*, 58 (5), 533-541.

Melhus, C.S. & Rivard, M.J. (2006). Approaches to calculating AAPM TG-43 brachytherapy dosimetry parameters for 137Cs, 125I, 192Ir, 103Pd, and 169Yb sources. *Med Phys*, 33 (6), 1729-1737.

Nath, R.; Anderson, L.L.; Luxton, G.; Weaver, K.A.; Williamson, J.F. & Meigooni, A.S. (1995). Dosimetry of interstitial brachytherapy sources: recommendations of the AAPM Radiation Therapy Committee Task Group No. 43. American Association of Physicists in Medicine. *Med Phys* 22 (2), 206-234.

Nucletron. (1998). General information. Netherlands.

Parsai, E.I.; Zhang, Z. & Feldmeier, J.J. (2009). A quantitative three-dimensional dose attenuation analysis around Fletcher-Suit-Delclos due to stainless steel tube for high-dose-rate brachytherapy by Monte Carlo calculations. *Brachytherapy*, 8 (3), 318-323.

Perez-Calatayud, J.; Granero, D. & Ballester, F. (2004). Phantom size in brachytherapy source dosimetric studies. *Med Phys*, 31 (7), 2075-2081.

Perez-Calatayud, J.; Granero, D.; Ballester, F. & Lliso, F. (2005). A Monte Carlo study of intersource effects in dome-type applicators loaded with LDR Cs-137 sources. *Radiother Oncol*, 77 (2), 216-219.

Perez-Calatayud, J.; Granero, D.; Ballester, F.; Puchades, V. & Casal, E. (2004). Monte Carlo dosimetric characterization of the Cs-137 selectron/LDR source: evaluation of applicator attenuation and superposition approximation effects. *Med Phys*, 31 (3), 493-499.

Rivard, M.J.; Butler, W.M.; DeWerd, L.A.; Huq, M.S.; Ibbott, G.S.; Meigooni, A.S. et al. (2007). Supplement to the 2004 update of the AAPM Task Group No. 43 Report. *Med Phys*, 34 (6), 2187-2205.

Rivard, M.J.; Coursey, B.M.; DeWerd, L.A.; Hanson, W.F.; Huq, M.S.; Ibbott, G.S. et al. (2004). Update of AAPM Task Group No. 43 Report: A revised AAPM protocol for brachytherapy dose calculations. *Med Phys*, 31 (3), 633-674.

Rivard, M.J.; Venselaar, J.L. & Beaulieu, L. (2009). The evolution of brachytherapy treatment planning. *Med Phys*, 36 (6), 2136-2153.

Sakelliou, L.; Sakellariou, K.; Sarigiannis, K.; Angelopoulos, A.; Perris, A. & Zarris, G. (1992). Dose rate distributions around 60Co, 137Cs, 198Au, 192Ir, 241Am, 125I (models 6702 and 6711) brachytherapy sources and the nuclide 99Tcm. *Phys. Med. Biol.*, 37, 1859–1872.

Sina, S. (2007). Simulation and Measurement of Dosimetric Parameters for ^{137}Cs Brachytherapy Source, Based on TG-43 Protocol by TLD and Monte Carlo. Shiraz univercity, Shiraz

Sina, S.; Faghihi, R.; Meigooni, A.S.; Mehdizadeh, S.; Mosleh Shirazi, M.A. & Zehtabian, M. (2011a). Impact of the vaginal applicator and dummy pellets on the dosimetry parameters of Cs-137 brachytherapy source. *Journal of Applied Clinical Medical Physics*, 12 (3), 183-193.

Sina, S.; Faghihi, R.; Meigooni, A.S.; Mehdizadeh, S.; Zehtabian, M. & Mosleh Shirazi, M.A. (2009). Simulation of the shielding effects of an applicator on the AAPM TG-43 parameters of CS-137 Selectron LDR brachytherapy sources. *Iran. J. Radiat. Res*, 7 (3), 135-140.

Sina, S.; Faghihi, R.; Zehtabian, M. & Mehdizadeh, S. (2011b). Investigation of anisotropy caused by cylinder applicator on dose distribution around Cs-137 brachytherapy source using MCNP4C code. *Iran. J. of Med. Physics*, 8 (2 (31)), 57-65.

Siwek, R.A.; Obrien, P.F. & Leung, P.M.K. (1991). Shielding effects of Selectron applicator and pellets on isodose Distributions. *Radiother. Oncol.*, 20, 132–138.

Song, G., & Wu, Y. (2008, Jun). In A Monte Carlo Interstitial Brachytherapy Study for AAPM TG-43 Dose Calculation Formalism in Heterogeneous Media. Paper presented at the 2nd International Conference on Bioinformatics and Biomedical Engineering, 2008. ICBBE 2008.

Tiourina, T.B.; Dries, W.J.F. & van der Linden, P.M. (1995). Measurements and calculations of the absorbed dose distribution around a 60Co source. *Med Phys*, 22 (5), 549–554.

UT-Battelle, & LLC. (2000). Rsic Computer Code Collection MCNP4C. New Mexico: Los Alamos National Laboratory.

Venselaar, J.L.M., Van der Giessen, P.H., & Dries, W.J.F. (1996). Measurement and calculation of the dose at large distances from brachytherapy sources: Cs-137, Ir-192, and Co-60. *Med. Phys.*, 23, 537-543.

Williamson, J.F. (1991). Comparison of measured and calculated dose rates in water near I-125 and Ir-192 seeds. *Med Phys*, 18 (4), 776-786.

Zehtabian, M.; Faghihi, R. & Sina, S. (2010). Comparison of dosimetry parameters of two commercially available Iodine brachytherapy seeds using Monte Carlo calculations. *Iran. J. Radiat. Res.*, 7 (4), 217-222.

Zehtabian, M.; Faghihi, R.; Sina, S. & Noorizadeh, A. (2010). Investigation of the Effects of Tissue Inhomogeneities on the Dosimetric Parameters of a Cs-137 Brachytherapy Source using the MCNP4C Code. *Iran. J. of Med. Physics*, 7 (3 (28)), 15-20.

2

Dose Calculations of the Ru/Rh-106 CCA and CCB Eyes Applicators

Itzhak Orion and Emanuel Rubin
Ben-Gurion University of the Negev, Israel

1. Introduction

One of the tumors that radiotherapy can be implemented is the choroidal melanoma - eye cancer. Since it is possible to bring the radiation source into the tumor vicinity, the use of a sealed beta radiation source became an applicable treatment. A short range of radiation of few millimetres or less with a minimal radiation spread to the surrounding can be preferable. Since the eye is a very sensitive organ high dose radiation could seriously damage it and even to cause blindness. The sensitivity of the eye parts is - in a descending order - the lens, corona, conjunctiva, retina, optic nerve, therefore it is important to map the accurate absorbed dose to the eye during a treatment (Egbert et al., 1980). Medium and large sized tumors are vastly treated with ^{125}I applicators, and β^- ray applicators such as ^{106}Ru are in use for small- sized tumors in eyes. ^{106}Ru ophthalmic applicators have been used for close to fifty years in the treatment of choroidal melanoma. Sixteen standard models of ^{106}Ru applicators are currently manufactured by BEBIG GmbH, Germany (BEBIG, 2003). The form of these applicators is a spherically concave silver bowl with an inner radius of curvature between 12 and 14 mm, and a total shell thickness of 1 mm. Various shapes with diameters between 11.5 and 25.5 mm are available.

The radioactive layer is electrically deposited with an approximate thickness of 0.1 mm on the concave surface of a 0.2 mm thick silver target foil. This target foil is, in turn, deposited between the concave surface of a 0.7 mm thick layer of silver (rear) and the convex surface of a 0.1 μm thick layer of silver (window). The precise applicator measures were provided by the manufacturer, BEBIG GmbH.

The ^{106}Ru parent disintegrates via beta decay with peak energy of 39 keV to a radionuclide daughter, the ^{106}Rh. The primary contributor to therapeutic dose is the continuous spectrum of beta particles emitted from the decay of ^{106}Rh (half-life ~30 sec). ^{106}Rh disintegrates by beta decay that its mean beta energy is of about 1.4 MeV and maximum of 3.54 MeV to ^{106}Pd (stable).

Two main papers on the subject of Monte Carlo simulations of the Ru/Rh-106 applicators were previously published. In the paper by Sánchez-Reyes et al. dose distribution results using the PENELOPE code (Salvat et al., 1996) were presented in 1998 (Sanchez-Reyes et al., 1998). The study using the PENELOPE code lead to different results than presented in this work, due to new developments in the electron transport model that were implemented after the year 2005. Šolc recently published simulation results of the MCNPX code for a COB-type applicator only, and showed a good agreement between the simulated doses to measurements results (Šolc, 2008).

Dose measurements were carried out using three-dimensional scintillation dosimetry for CCX-type eye applicators, and isodose maps were obtained (Kirov et al. 2005). The measurements results were presented for 4 mm depth and higher only.

A high accuracy model for the simulations of the CCA-type and the CCB-type applicators in the one of the most updated Monte Carlo code for electrons, the EGS5, is presented in this chapter. The choice of the applicator type is made by the physician according to the size and depth of the treated tumor.

2. Materials and methods

The applicators are made of the ^{106}Ru isotope with half life of 373.59 days with maximum energy of 39 keV. The 39 keV beta rays of the Ru cannot escape the applicator window due to their short range in silver. In the simulation, therefore, the beta rays of the ^{106}Ru were not taken into account. The decay formulae are listed in Eq. (1). The daughter ^{106}Rh emits a continuum spectrum of beta of the following main energies: 1.51 MeV (79%), 0.97 MeV (9.7%), 1.27 MeV (8.4%). The ^{106}Rh and the ^{106}Ru properties are listed in Table 1. The overall continuum beta emission of the ^{106}Rh, shown in Figure 1, was fitted and programmed as an energy source distribution to the EGS user-code.

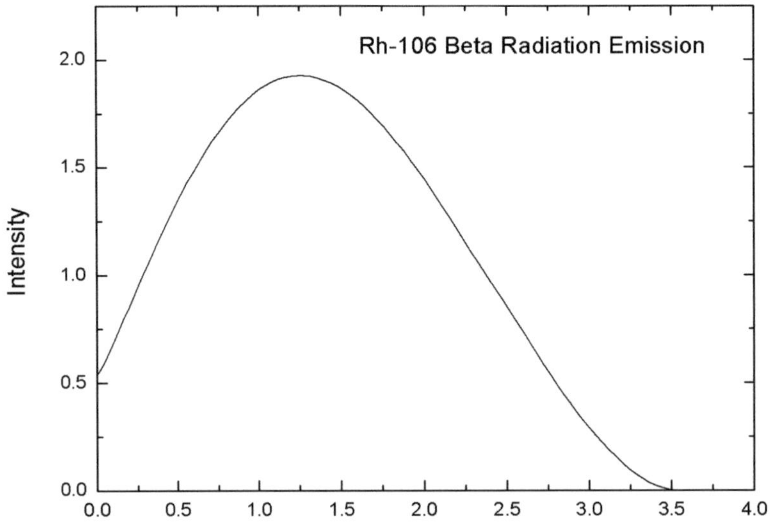

Fig. 1. The Ru-106 total beta emission spectrum used in the simulations [data was taken from JEF-PC 2.0 database (Konieczny , 1997)].

$$^{106}_{44}Ru \rightarrow ^{106}_{45}Rh + \beta^- + \bar{\nu}$$

$$^{106}_{45}Rh \rightarrow ^{106}_{46}Pd + \beta^- + \bar{\nu} \qquad (1)$$

Radioisotope	^{106}**Rh**	^{106}**Ru**
Atomic number	45	44
Atomic mass	106	106
Half life	29.80 sec	373.15 day
Q-value (keV)	3541	39.40
Decay	β^-	β^-
Density (g cm^{-3})	12.4	12.2

Table 1. The beta source properties (From ENSDF - NNDC).

In Figs 2,3 the vertical cross-sectional view of each applicator are shown. The CCA-type applicator dimensions are: concave with a radius of 12.0 mm; Height h = 3.3 mm; The whole applicator diameter is 15.3 mm; The radioactive part diameter is 13.4 mm. Total applicator angle of 72.2° (2 x 36.1°); The radioactive part angle of 67.4° (2 x 33.7°).

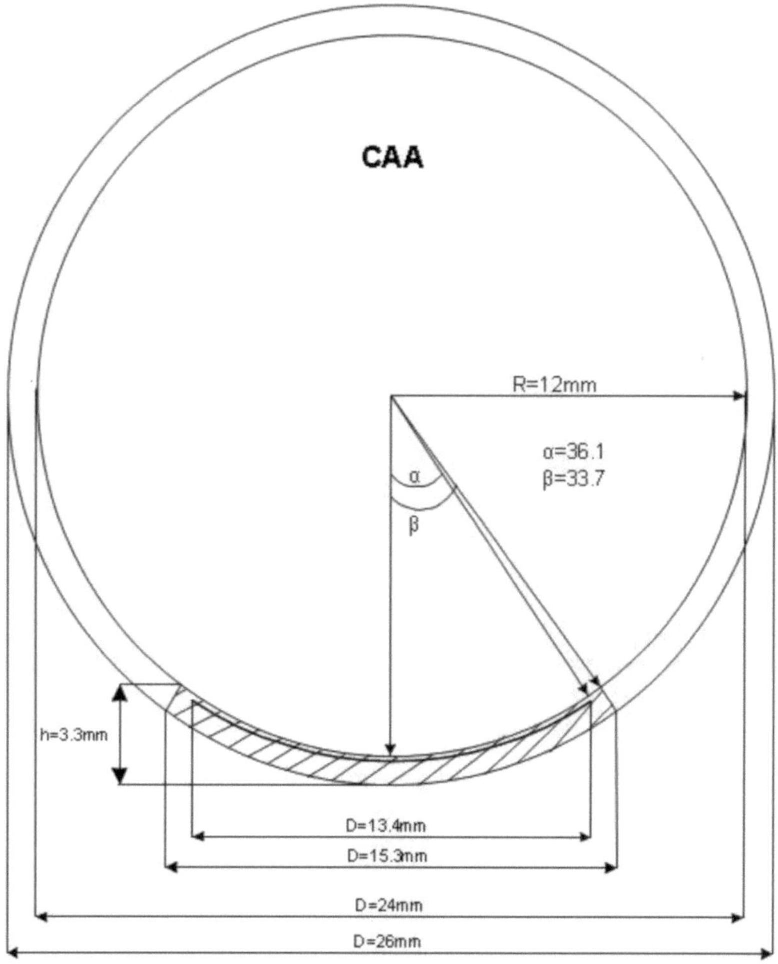

Fig. 2. The vertical cross-sectional view of CCA-type applicator with dimensions.

The CCB-type applicator dimensions are: concave with a radius of 12.0 mm; Height h = 5.4 mm; The whole applicator diameter is 20.0 mm; The radioactive part diameter is 18.2 mm. Total applicator angle of 101.8° (2 x 50.9°); The radioactive part angle of 97.8° (2 x 48.9°).

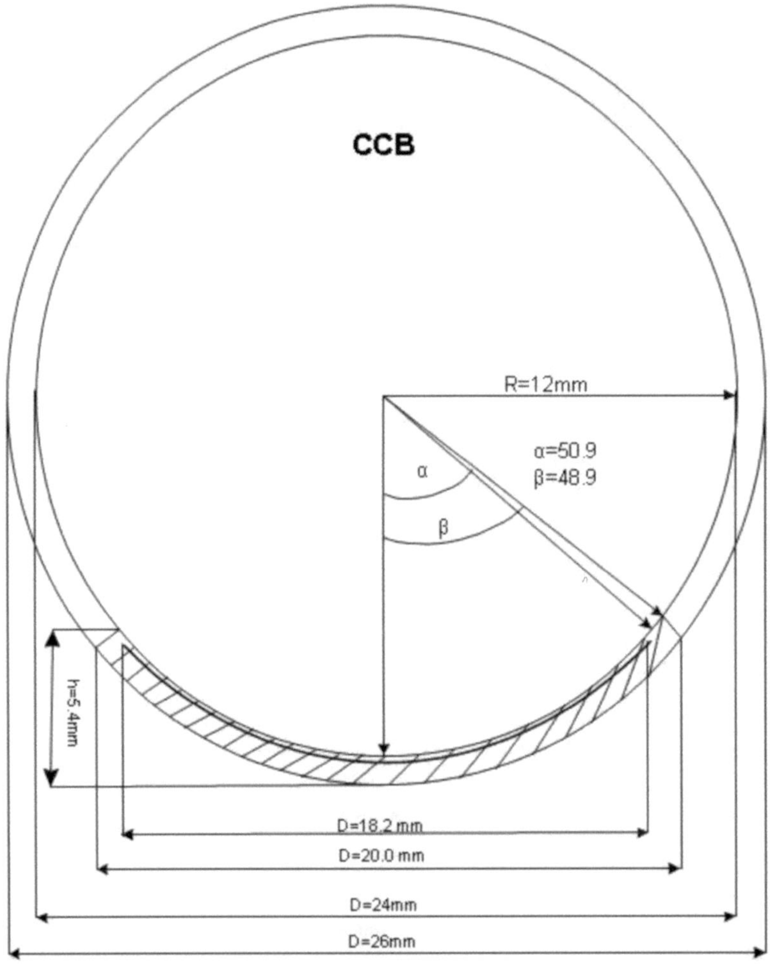

Fig. 3. The vertical cross-sectional view of CCB-type applicator with dimensions.

2.1 The beta emission area distribution

In order to distribute the beta emission source area on a concave surface, PDF (Probability Density Functions) were formulated using spherical functions. The total areas for the CCA-type and for the CCB-type are:

$$Total\ Area_{CCA} = r^2 \int_0^{2\pi} d\theta \int_0^{\frac{33.7\cdot\pi}{180}} \sin(\varphi)\, d\varphi = 0.336\pi r^2 \tag{2}$$

$$Total\ Area_{CCB} = r^2 \int_0^{2\pi} d\theta \int_0^{\frac{48.9\cdot\pi}{180}} \sin(\varphi)\, d\varphi = 0.543\pi r^2 \tag{3}$$

Hence, after taking into account each case integration limits, the specific cumulative density functions (CDFs) are:

$$\begin{aligned} F(\theta,\varphi) &= r^2 \zeta_1 \theta\left(1-\cos(\zeta_2 \varphi)\right) \\ F(\theta) &= 2\pi\zeta_1 \\ F(\varphi)_{CCA} &= \frac{180}{33.7}\arccos(\zeta_2) \\ F(\varphi)_{CCB} &= \frac{180}{48.9}\arccos(\zeta_2) \end{aligned} \tag{4}$$

Where ζ notes a uniform random number in the range [0,1].

Each of the source positions on the surface were set to introduce an isotropic emission probabilities.

2.2 The Monte Carlo simulation code

The Monte Carlo method provides approximate solutions to a variety of mathematical problems by performing statistical sampling that rely on repeated random sampling. A computer calculates the results of simulated experiments. The method is used to resolve problems with no probabilistic content as well as those with inherent probabilistic structure, such as the interaction of nuclear particles with materials. It is particularly useful for complex problems that cannot be modelled by computer codes that use deterministic methods.

The EGS (Electron-Gamma Shower) code system is a general purpose package for the Monte Carlo simulation of the coupled transport of electrons and photons in an arbitrary geometry for particles with energies ranging from above a few keV up to several hundred GeV. The EGS5 is a FORTRAN open source program. Since the 1990s when the previous EGS4 code was released, it has been used in a wide variety of applications, particularly in medical physics, radiation measurement studies, and industrial development. The EGS5 code system (Hirayama et al. 2006) contains, among many other subprograms, four user-called subroutines: BLOCK SET, PEGS5, HATCH, and SHOWER. These routines call other subroutines in the EGS5 code, some of which call two user-written subroutines, HOWFAR and AUSGAB, which respectively define geometry, and scoring output. The EGS5 transport code for electrons is fundamentally different and advanced from the previous EGS4 code. The electron step in EGS5 is treated by splitting each step into two segments, and a scattering hinge is applied in between the segments (Bielajew & Wilderman, 2000). The user communicates with EGS5 by means of the subroutines mentioned above which enable him to access variables contained in various COMMON blocks. To use EGS5, the user must write a MAIN program and the subroutines HOWFAR and AUSGAB.

The Monte Carlo simulations were written using the EGS5 code system. The EGS5 Monte Carlo code system is a new generation of the EGS4 well validated code for photons and electrons transport. The EGS5 consists of newly developed pseudo-random generator, accurate low-energy photon fluorescence transitions and new algorithm for Bremsstrahlung radiation, which assure reliable results for radiation simulations studies. The EGS5 includes several tools, such as the CG-VIEW for geometry editing and as a viewer, and the PEGS5 editor for media cross sections definitions. All kinetic energy cut-offs were set to be at 1 keV along this study. The EGS5 is running under LINUX operating systems. A schematic flowchart to illustrate how the user is able to prepare a simulation in EGS5 is presented in Figure 4.

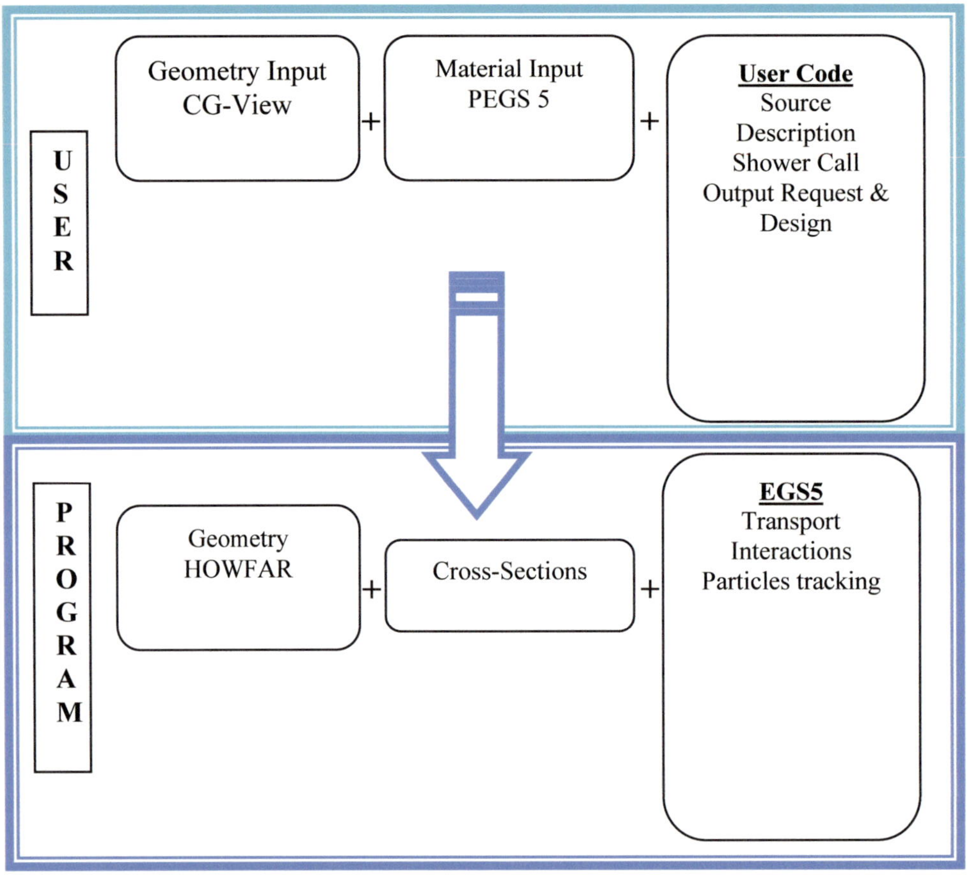

Fig. 4. The EGS5 code system schematic layout.

The detailed geometry of each applicator as described above was input and checked for the simulations using the CG-VIEW program, which is a complementary part of the EGS5 package. The media used in the simulations are: Ag and Ru for the applicator, soft tissue [based on ICRU44: tissue, soft ICRU Four-Component (ICRU – 44, 1989)] as the filling of the eye.

The CCA-type applicator simulation was visually inspected by tallying 200 primary electron tracks into the CG-VIEW program. The electron tracks results are shown in Figure 5 (the secondary photons lines were not represented). The basic structure of the electrons range and spread in the applicator and in the eye can be seen.

The simulations tallied energy deposition in 1 mm diameter spheres in order to calculate the absorbed dose. Each run, consist of 5 million histories, took about 25 hours on a PC computer (Intel® Pentium ® 1.60 GHz).

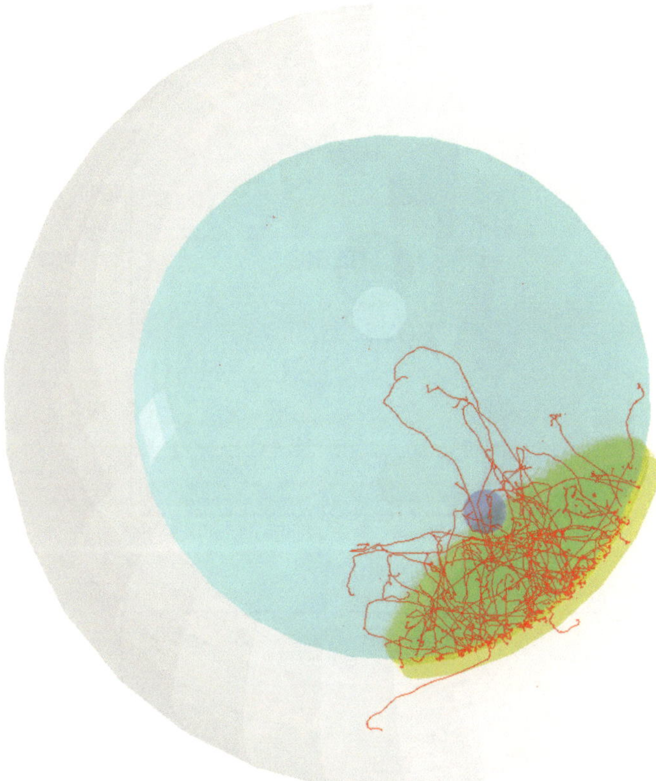

Fig. 5. Electron tracks from the simulated CCA-type applicator as presented by the CG-VIEW program.

3. Results and discussion

The results provided the dose delivered in a treatment of choroidal melanoma. The EGS5 depth dose results were compared to the BEBIG manufacturer results that were supplied with the applicators data sheets. The measured results were obtained from a 1 mm x 0.5 mm scintillator with an error of about 20 % (Fluhs et al., 1996). The measured results and the simulations results were both normalized to be 100 at 2 mm depth. The comparison for each applicator type is shown in Figure 6. The comparison showed mostly a good agreement between the measurements and the simulations results. The simulations statistical uncertainties were analyzed to be discussed in details in the conclusions section.

The results of the radial dose distributions in Figure 7 were obtained using another set of EGS5 Monte Carlo simulations. The absorbed dose was accumulated in 1 mm radius spherical unit-cells positioned around the eyeball center at a 11 mm radius across a perpendicular plan to the applicator. The 0° angle is at the closest position to the applicator concave center, where the dose was normalized to be of a value of 100. The radial dose for the CCA-type and for the CCB-type applicators showed a decrease of five orders of magnitude along a range of 120° (Figure 8).

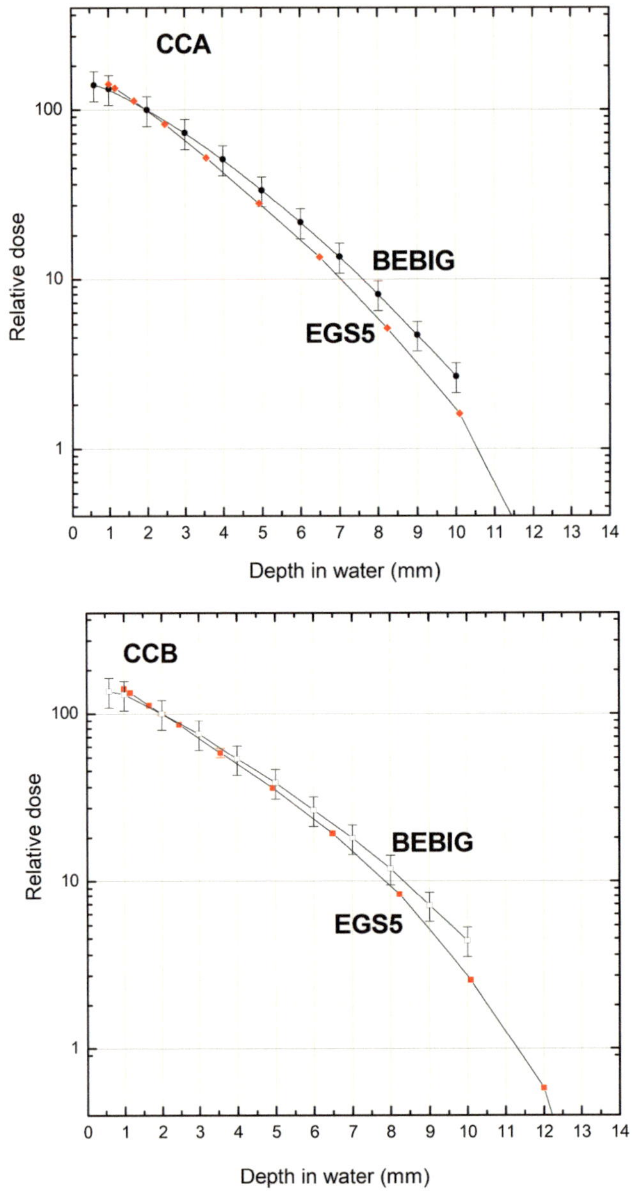

Fig. 6. Comparison of the dose versus depth in water measurements to simulations results for the CCA-type and for the CCB-type applicators.

The normalized absorbed dose (100 at 2 mm distance) along several plans at 2 mm to 8 mm above the applicators were obtained from the simulations, and plotted in Figure 7. The CCA-type dose is more flat even at lower plans compared to the CCB-type dose.

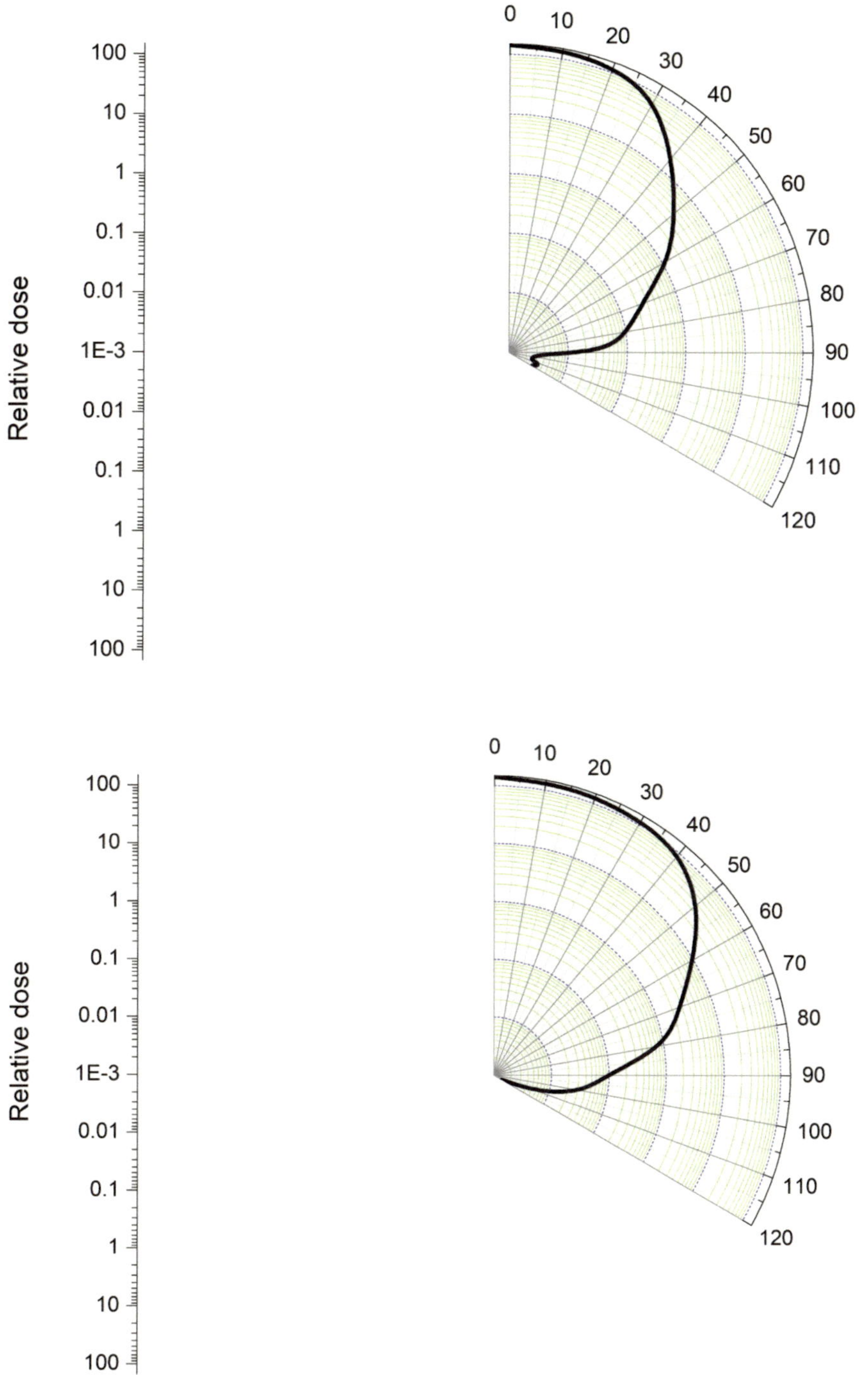

Fig. 7. The radial dose distributions in a 1 mm radius unit-cell moving around the eyeball center at a 5 mm radius across a perpendicular plane to the applicator.

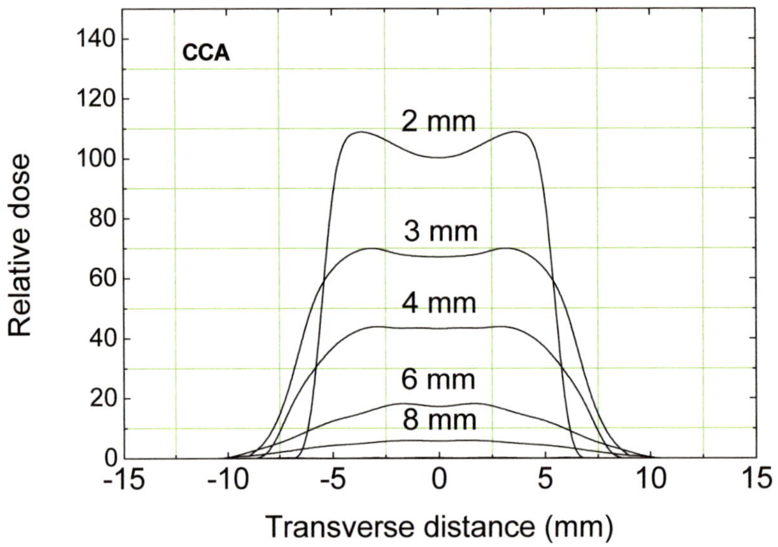

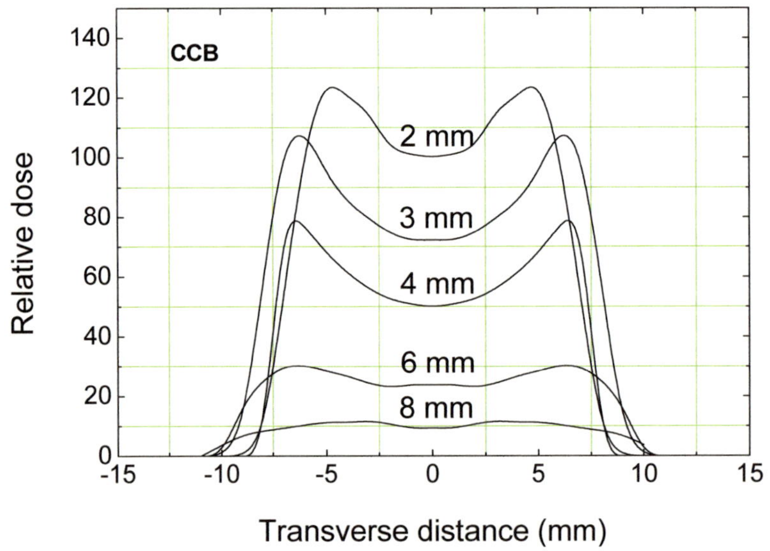

Fig. 8. The EGS5 normalized absorbed dose along several planes at 2 mm to 8 mm above the applicators.

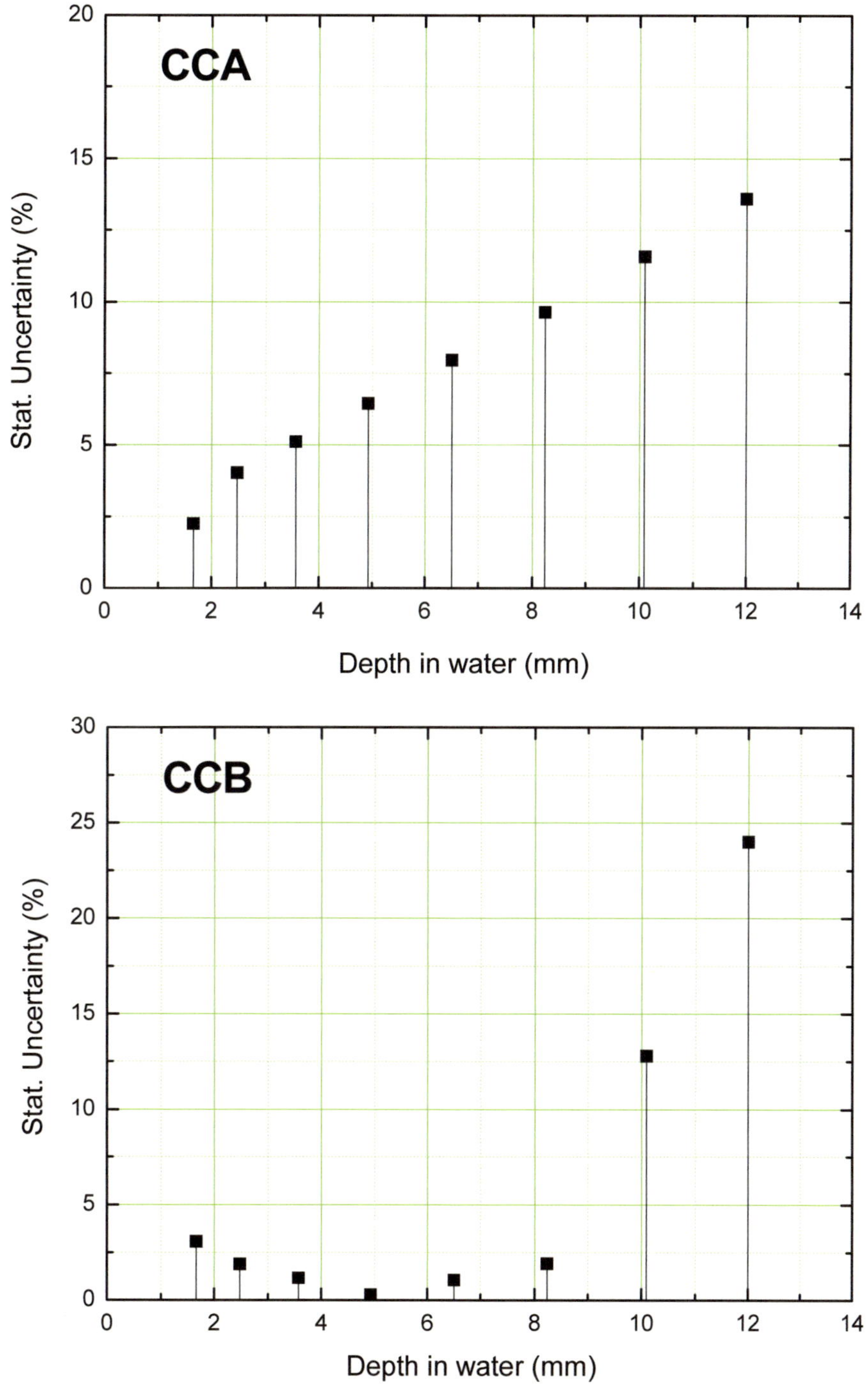

Fig. 9. The simulations statistical uncertainties versus depth in water for CCA-type and for CCB-type applicators.

Depth (mm)	EGS5 Dose (norm. at 2 mm)	PENELOPE Dose (norm. at 2 mm)	Difference (%)
1.0	142.1	144.7	1.8
1.5	113.0	117.0	3.5
2.0	100.0	100.0	0.0
2.5	82.3	78.7	4.4
3.0	67.7	74.5	9.5
3.5	52.1	57.4	9.8
4.0	44.5	40.4	9.5
5.0	27.1	25.5	5.8
6.0	18.3	17.0	7.4
6.5	13.5	12.8	5.6
7.0	10.7	10.6	0.7
8.0	6.8	6.8	0.1

Table 2. Comparison of the EGS5 simulation depth dose results from this study to PENELOPE results after normalization (taking from Sanchez-Reyes, 1998): a) CCA-type Applicator.

Depth (mm)	EGS5 Dose (norm. at 2 mm)	PENELOPE Dose (norm. at 2 mm)	Difference (%)
1.0	140.9	131.9	6.6
1.5	111.7	114.9	2.8
2.0	100.0	100.0	0.0
2.5	85.9	89.4	3.9
3.0	72.6	78.7	8.1
3.5	58.3	70.2	18.6
4.0	51.2	61.7	18.6
5.0	34.9	46.8	29.2
6.0	24.8	40.4	48.0
6.5	19.2	29.8	43.3
7.0	15.6	25.5	48.4
8.0	10.4	17.0	47.9

Table 3. Comparison of the EGS5 simulation depth dose results from this study to PENELOPE results after normalization (taking from Sanchez-Reyes, 1998): b) CCB-type Applicator.

3. Conclusion

The dose calculations were carried out for two main applicator types, CCA-type and CCB-type, using the EGS5. The Monte Carlo simulations results obtained in this study are close to the manufacturer's measured data. Comparison to measurements showed a difference of about 10% (CCB) and 20% (CCA) at 5 mm depth. The dose statistical uncertainties in the calculations show that for an amount of 10% for the CCA-type and an uncertainty of 2.5%

for the CCB-type, at 8.5 mm depth. However, in both applicators the dose reduction was found over one magnitude up to a distance of 7.5 mm.

The simulation statistical uncertainties versus depth are shown in Figure 9. Different depth dependence was found for CCA-type applicator compared to CCB-type applicator. The CCA-type uncertainties are constantly increasing with depth, whilst the CCB-type statistical uncertainties were not gradually increased.

This different behavior can be explain by the nature of the active area radioactive distribution, when in CCB-type the angle of the active area is much vast resulting a higher flux toward the central axis up to a distance of 8.5 mm. The EGS5 statistical uncertainties assessments showed that the CCB-type source might be proffered, due to low uncertainties along the treatment range.

Since the same applicators dose was simulated in a previous Monte Carlo simulations study using a different Monte Carlo code, the PENELOPE (Sanchez-Reyes et al., 1998), it was interesting to examine the difference of the dose results. The relative dose versus depth in water from the EGS5 results was compared to the relative dose from PENELOPE after normalizing the values to be 100 at 2 mm depth. The comparison was preformed for the CCA-type and for the CCB-type applicators, as listed in Tables 2,3. Even the CCA-type applicator's external diameter of the active area is different from the current work (15.5 mm instead of 15.3 mm), the results showed up to 10 % only dose difference that can be explained by the statistical uncertainty. The CCB-type comparison results showed increasing dose difference toward 49 % along the 8 mm depth in water. The EGS5 showed always lower dose compared to the PENELOPE results in the CCB-type applicator, which is might be expressed due to the different electron transport of the EGS5 code. The CCA-type applicator results of the EGS5 study showed transverse dose profile that has a clear shape difference in Figure 6 compared to the PENELOPE published results [(Sanchez-Reyes et al., 1998) see Figure 4]. It can be seen that EGS5 could follow fine profile structure of the CCA-type, while PENELOPE could result that fine structure in the CCB-type that causes a more non smooth shape.

This study showed the potential of using Monte Carlo simulations in order to calculate the radiation dose delivered to the eye in high accuracy in a short computing time, which may assist the choice of applicator type for every individual treatment.

4. Acknowledgment

The authors would like to thank Professor Yaakov Peer from Hadassah University Hospital in Jerusalem, Israel. We thank BEBIG Company for the applicators data and details supplied that assist to perform the research.

5. References

BEBIG; (2003). Ruthenium eye applicator customer information: introduction of the new NIST-calibrated dosimetry introduction of the new PTB - calibrated activity measurement. Berlin.
Available from http://www.bebig.eu/products/ophthalmic.html

Bielajew A. F. & Wilderman S. J. (2000). Innovative Electron Transport Methods in EGS5, Proceedings of the 2nd International Workshop on EGS, KEK, Japan (KEK Proceedings 2000-20).

Egbert P. R.; Fajardo L.F.; Donaldson S. S. & Moazed K. (1980). Posterior ocular abnormalities after irradiation for retinoblastoma: a histopathological study, *Br. J. Ophthalmol*, Vol. 64 pp. 660-665

ENSDF, "Evaluated Nuclear Structure Data File" National Nuclear Data Center, Brookhaven National Laboratory, Upton, NY 11973-5000
Available from http://www.nndc.bnl.gov/

Fluhs D.; Heintz M.; Indenkampen F.; Wieczorek C.; Kolanoski H. & Quast U. (1996). Direct reading measurement of absorbed dose with plastic scintillators—the general concept and applications to ophthalmic plaque dosimetry, *Medical Physics* Vol. 23, pp. 427–434

Hirayama H.; Namito Y.; Bielajew A. F.; Wilderman S. J. & Nelson W. R. (2006). *The EGS5 code system* SLAC-R-730 / KEK Report 2005-8

ICRU - 44 (1989). *Tissue Substitutes in Radiation Dosimetry and Measurement*, Report 44 of the International Commission on Radiation Units and Measurements, Table 2 (Bethesda, MD)

Kirov A.S.; et al. (2005). The three-dimensional scintillation dosimetry method: test for a ^{106}Ru eye plaque applicator, *Phys Med Biol.* Jul 7; 50(13):3063-81

Konieczny M. et al. (1997). JEF-PC Version 2.0, *Proc. Int. Conf. on Nuclear Data for Science and Technology*, Trieste, Italy, May 1997, p. 1063

Salvat F.; FernaÂndez-Varea J. M.; Baro Â. J. & Sempau J (1996). *PENELOPE, an algorithm and computer code for Monte Carlo simulation of electron photon showers*, CIEMAT Report n. 799. Madrid

Sanchez-Reyes A.; Tello J. J.; Guix B. & Salvat F. (1998). Monte Carlo calculation of the dose distributions of two ^{106}Ru eye applicators, *Radiotherapy and Oncology* 49, pp. 191-196

Šolc J. (2008). Monte Carlo calculation of the dose water of a COB-type ophthalmic plaque, Third McGill International Workshop, *Journal of Physics: Conference Series 102*

Section 2

Californium

3

Progress in Californium-252 Neutron Brachytherapy

C.-K. Chris Wang
Georgia Institute of Technology, USA

1. Introduction

The potential of using Cf-252 to treat cancer patients was first described by Shlea and Stoddard in 1965 as a source of neutrons using brachytherapy procedures [Stoddard, 1986]. In the late 1960's and early 1970's, the United States and England initiated human clinical trials of Cf-252 **n**eutron **b**rachy**t**herapy (NBT) for treating patients of various cancer types. These earlier trials treated only very few patients and were given up prematurely due to then emerging interest in accelerator-based **e**xternal **b**eam **f**ast **n**eutron **t**herapy (EBFNT).

In 1976, Dr. Maruyama, then at the University of Kentucky Hospital, began the NBT clinical trials with a focus on intracavitary treatment of advanced cervix and GYN cancers. During the fifteen year period (1976-1991) at Kentucky, Dr. Maruyama had treated several hundred cervical cancer patients and obtained successful results [Maruyama et al., 1991]. Based on the successful cervical cancer trials at Kentucky, Dr. Maruyama moved to the Harper Hospital of Wayne State University and intended to expand the NBT trials to include interstitial treatment of many other cancer types. The clinical trials at Harper, however, did not last long after Maruyama's unexpected death in 1995.

In parallel with the effort of Dr. Maruyama, several groups in Russia, Czech Republic, and Japan began studies and clinical trials. These groups met several times in USA in 1985, 1990, and 1997 and exchanged their experiences with NBT. Great efficacy has been found for cervix cancers of all stages but most notable for the advanced stage III and bulky IB cancers [Maruyama et al., 1997; Tacev, et al. 2003]. Endometrial adenocarcinomas and vaginal cancers [Maruyama et al., 1997] were also very curable with NBT. Various oral cavity cancers, recurrent tumors, radioresistant tumors such as melanoma, sarcoma, and glioblastoma were effectively treated by NBT [Vtyurin & Tsyb, 1986; Tsuya & Kaneta, 1986; Maruyama & Patel, 1991; Medvsdev et al., 1991; Stoll et al., 1991; Vtyurin et al., 1991a; Vtyurin et al., 1991b]. It was estimated by 1997 that approximately 5,000 patients have been treated with NBT. In 1999 China implemented its first NBT afterloading system (using Russian-made Cf-252 sources) to treat mainly cervical cancers. Presently 18 NBT afterloading systems have been installed in China, with approximately 20,000 patients treated. A latest report, based on the follow-up data of 696 cervical cancer patients treated with NBT, shows that the overall survival rate at 3 and 5 years are clearly better and that the late complications are lower than those treated with the conventional gamma-brachytherapy

[Lei et al., 2011]. While there is no standard treatment regimen exists for NBT, an overwhelming majority of the clinical studies were based on a mixed treatment method of which NBT is implemented first and then followed with the traditional external-beam radiotherapy (EBRT) .

While much progress has been made in intracavitary NBT over the years, currently there is no clinical implementation in interstitial NBT. The major obstacle for interstitial NBT has been that either the size of the Cf-252 source was too big or the activity was too low. In a review article [Maruyama et al., 1997], which was published after his death, Dr. Maruyama stressed the need for developing small size high activity Cf-252 source seeds and afterloading machines to facilitate clinical trials on the interstitial treatment of brain tumors, sarcomas, and a large variety of tumors. With this mission in mind, Isotron Inc. (a start-up company in Alpharetta, Georgia, USA) in 1999 entered a 5-year corporative research and development agreement (CRADA) with the Oak Ridge National Laboratory (ORNL) of the U.S. to develop a new generation high activity miniature source seeds. In October 2002, under the CRADA, ORNL/Isotron successfully encapsulated a batch of new Cf-252 source seeds with size and activities suitable for interstitial NBT [Martin, 2002]. The overall effort of Isotron of bringing interstitial NBT to full commercial stage, unfortunately, did not reach fruition due to financial difficulties.

While the methodology of ^{252}Cf-based NBT is the same as that of the conventional brachytherapy modalities (based on gamma-emitting isotopes such as Cs-137, Ir-192, I-125, Pd-103, etc.), i.e. to bring the source close to (or into) the tumor volume of a patient, the characteristic of neutron emissions and the associated radiobiological effect make the NBT unique.

2. Califronium-252 production and availability

The isotope of Cf-252 was first discovered in 1952 in the debris from uranium that had been subjected to intense neutron irradiation [Fields, 1956]. Early investigation showed that Cf-252 has a half-life of between 2 and 3 years and a significant branching fraction for decay by spontaneous fission, making it an especially compact source of neutrons. These desirable properties led to a sustained national effort of the U.S. to produce and recover macroscopic quantities of Cf-252. This effort began in late 1952 with irradiation of multigram quantities of plutonium-239 (or Pu-239) in the Materials Test Reactor (MTR) at the Idaho National Engineering Laboratory and culminated in the eventual recovery of purified microgram quantities of Cf-252 in 1958 at Lawrence Berkeley Laboratory [Mosley et al., 1972].

As the scientific interest in Cf-252 grew, the demand for the isotope quickly exceeded the supply. As a result, a National Transplutonium Element Production Program was undertaken to produce large quantities of Cf-252 and other transplutonium isotopes for the research community. This new program led to a large scale market evaluation program at the Savannah River Plant (SRP) and a smaller research effort at Oak Ridge National Laboratory (ORNL) beginning in the late-1960s. Since 1973, the western world's supply of Cf-252 (which is about two-thirds of the world's supply) is produced in the High Flux Isotope Reactor (HFIR) and recovered at the Radiochemical Engineering Development Center (REDC) at ORNL [Martin et al., 1997]. The rest one-third is produced at Russia's Research Institute of Atomic Reactors facility, and is used mainly by Russia and China.

Initially, Cf-252 was produced at HFIR/REDC from the neutron irradiation of Pu-242. By 1968, most of the Pu-242 had been irradiated and transmuted to curium-244 (or Cm-244). Continual irradiation and recovery of Cm-244 led to subsequent buildup of Cm-246 and Cm-248, which greatly enhances the yield of Cf-252. This is because fewer neutrons are required to produce Cf-252. A Cf-252 production cycle involves the irradiation of 11 to 13 targets for 6 to 8 months in HFIR, followed by four months of processing and recovery in the remotely operated and maintained hot cells at REDC. A recent production cycle completed at HIFR/REDC yields typically 200-300 milligrams of Cf-252.

In May 2008, the US department of energy (DOE) announced that it was terminating its transuranic waste research and defense programs, putting the continued supply of Cf-252 to the western world in question. In May 2009, industry users and Cf-252 source manufacturers reached agreements with DOE to privately fund the Cf-252 program through FY 2012 with expectations that production will continue beyond that year and that future irradiation campaigns will be conducted every two years [NIDC, 2011].

3. Evolution of medical Cf-252 source fabrication

3.1 Earlier sources fabricated at SRL

The medical fabrication techniques were first developed at Savannah River Laboratory (SRL) [Mosley et al., 1972]. In general, operations are conducted remotely in a heavily shielded hot cell. Palladium (or Pd) is deposited onto a fine precipitate of californium oxalate, $Cf_2(C_2O_4)_3$, in an aqueous solution. The Pd-coated particles are then dried, calcined to Pd-coated Cf_2O_3, pressed into a pellet of approximately 50% theoretical density, sintered to 1300 °C, pressed again to approximately 90% theoretical density, then pressed into a capsule of platinum-10% iridium alloy, and sealed as a billet. At SRL, draw dies were used to reduce the diameter of this billet to as small as 0.30 mm in diameter.

One or more of the encapsulated cermet wires produced at SRL were included in a much larger source, the Applicator Tube (AT), which was used by Dr. Maruyama and collaborators at the University of Kentucky for intracavitary (e.g. gynecological) treatment. Figure 1 shows the doubly encapsulated AT source assembly. As shown, the outside diameter and length are 2.8 mm and 23.1 mm, respectively. A newly made AT source typically contained 10 to 30 μg of Cf-252. As the AT sources are obviously too large to be used for interstitial treatment, smaller seed capsules (< 1.1 mm dia.) had also been made at SRL. These seed sources, however, only contained 0.5 μg of Cf-252, a quantity that is too small for any practical use [Mosley et al., 1972]. Accordingly, in his review article [Maruyama et al., 1997], Dr. Maruyama stressed the need for developing small size high activity seed sources and afterloading machines to facilitate clinical trials on the interstitial treatment of brain tumors, sarcomas, and a large variety of tumors.

The current NBT treatment in China is implemented via an afterloading system (referred as Neutron Knife) developed by Zurua Science & Technology Co. As mentioned, a total of 18 systems have been deployed. According to the information available on the website of Zurua, each newly loaded source typically contains up to 1 mg of Cf-252 provided from Russia. While the activities of these sources are significantly higher than that of the AT sources, the dimensions (3 mm in outside diameter and 11 mm in length) are much too big, and therefore, the Neutron Knife system is only suitable for intracavitary NBT.

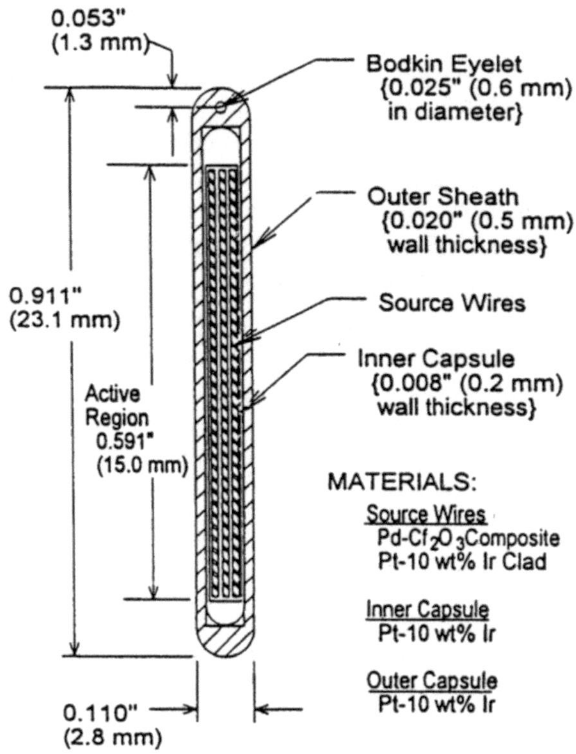

Fig. 1. The doubly encapsulated application tube (AT) source assembly

3.2 New high-activity miniature sources

It is worth noting that the active Pd-Cf_2O_3 "cermet" material included in the AT sources at SRL contains less than 0.1 wt% Cf-252. The REDC at ORNL have modified the SRL techniques to increase the Cf-252 content in the cermet material. Instead of sintering at 1300 °C, the pellet is heated to 1600 °C, which melts the Pd-Cf_2O_3 mixture. After cooling, the melted pellet is sufficiently strong and malleable to roll into a thin wire on a jeweler's rolling mill. For commercial sales, 1.1 mm square cermet wires are routinely fabricated with a nominal loading of 500 μg Cf-252 per inch (>0.1 wt% Cf-252) [Martin et al., 1997].

While the new procedure significantly increases the Cf-252 content (or specific activity) of the cermet wire, the high concentration of californium oxide necessarily degrades Pd workability. As a result, it was not possible to roll the cermet into a diameter below 0.6 mm, the size needed to make sources for interstitial treatment. In 1999, under a CRADA with Isotron, Inc., the ORNL developed a new wire shaping method that uses a modified swaging technique. That is, the wire is fed through a "shaper" unit, in which pneumatically activated collets hammer the circumference of the wire [ORNL, 2002]. A wire diameter of < 0.6 mm with uniform cross-section was obtained after repeated hammering at pressures between 40 and 100 psia. In October, 2002, a batch of seven high activity miniature source seeds was successfully encapsulated (Figure 2). The outside dimensions of the source capsule are 1.1 mm in diameter and 8 mm in length. The active length of the source is 5 mm. It is singly encapsulated with the wall thickness of 0.2 mm. The wall is made of platimum-

10% iridium alloy. The average quantity of Cf-252 in each source seed was approximately 90 μg, which is 200 times of that in the old SRL sources of a comparable size. The single wall and the small wall thickness cause the concern of structural integrity of the capsule wall. Specifically, the concern is about whether or not the capsule wall is strong enough to withstand the pressure build-up due to helium production from alpha decays [Rivard, 2000]. As of this writing, which has been more than three half-lives passed since the encapsulation, all seven sources show no compromise of their structural integrity. The small size of these new sources allow them to be used with remote high-dose-rate (HDR) afterloading systems comparable to current ones already in use for ^{192}Ir interstitial gamma brachytherapy. This new generation of source seeds thus made interstitial NBT practical.

One of the new sources was shipped to Georgia Institute of technology and National Institute of Standard Technology (NIST), where measurements were made on both neutron and gamma-ray dose profiles as well as the absolute total neutron intensity. Detailed discussion on the measured results associated with the new source is provided in Section 6. As of this writing, no commercial HDR unit has been built employing the new source for NBT. As mentioned in Section 1, all the HDR units developed in China employ the much larger Cf-252 sources supplied from Russia.

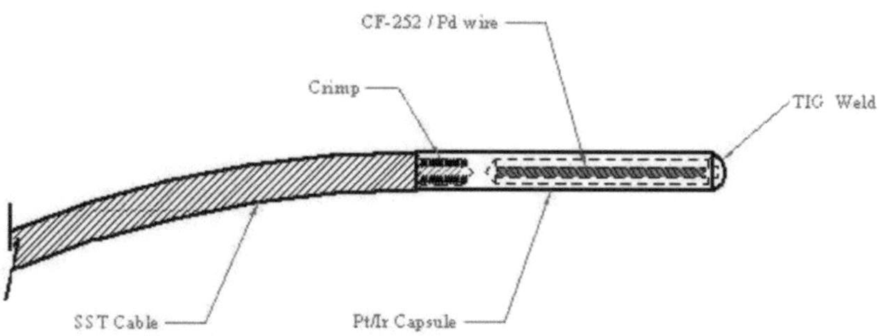

Fig. 2. The singly encapsulated new miniature source assembly

4. Biological basis of Cf-252 neutron brachytherapy

4.1 Neutrons and gamma rays and their interactions with tissue

Neutrons and gamma rays are often referred to as "indirect" ionizing radiations. Ionizing radiations consist of particles that are energetic enough to produce ions, break chemical bonds and alter biological function. Exposure to such radiation, therefore, can cause molecular and cellular changes such as mutations and chromosome aberrations, which in turn, may lead to cell death. Energetic charged particles (e.g. electrons, protons, alpha particles, or carbon ions), are referred to as "direct" ionizing radiations because they directly transfer their kinetic energies to a medium via Coulumb interactions with the atoms of the medium and cause ionizations (and excitations). Neutron and gamma rays are referred as "indirect" ionizing radiations because they must first interact with a medium to produce energetic charged particles, which will then cause ionizations/excitations in the medium. Gamma rays produce energetic charged particles (i.e. electrons) in a medium via atomic interactions such as photoelectric absorption, Compton scattering, or pair production. Likewise, neutrons produce energetic charged particles in a medium via nuclear interactions such as elastic scattering with the atomic nuclei, (n,p), and (n,α) reactions.

The radioisotopes that are commonly used in radiotherapy are mostly gamma emitters. These isotopes include Ra-226, Cs-137, Co-60, Ir-192, I-125, Pd-103, etc. Californium-252 is the only neutron-emitting radioisotope that has been used in radiotherapy. However, Cf-252 emits more than just neutrons. It also emits gamma rays, alpha particles, and beta particles (i.e. electrons). Because alpha particles and beta particles have very short ranges (< 1 mm), they usually do not penetrate the capsule wall of the source. As such, in a Cf-252 brachytherapy treatment, only neutrons and gamma rays emerge and interact with the surrounding tissue. As a rule of thumb, one third of the absorbed radiation dose in tissue surrounding a Cf-252 brachytherapy source is due to gamma rays and two thirds of the dose is due to neutrons [Rivard, 1999]. However, because neutrons (on the per unit dose basis) in average are a factor of 6 more effective in cell killing than the gamma rays, the biological effect of a Cf-252 brachytherapy treatment is overwhelmingly attributed to neutrons. The following subsection explains why neutrons are more effective in cell killing than gamma rays and how the biological effectiveness of Cf-252 neutrons is quantitatively assessed.

4.2 Relative Biological Effectiveness (RBE) of high-LET radiations

In general, the sub-cellular damages that bear biological consequences (e.g. cell death or transformation) result from densely packed ionization and excitation events taking place near or on a DNA molecule (or chromatin fiber). Sparsely distributed ionization/excitation events do cause DNA damages. But these damages are mostly mild and can be quickly repaired by the inherent enzyme-driven mechanisms of the cell. As such, one often relates the biological effectiveness of an ionizing radiation to a quantity called **l**inear **e**nergy **t**ransfer (LET), defined as the average amount of energy that is lost in a microscopic distance, e.g. one micrometer, approximately the size of a chromosome inside a cell nucleus. Gamma/x-rays and electrons are referred to as low-LET radiation because they produce sparsely distributed ionization/excitation events that give a low LET value (~0.3 keV/μm). Heavy ions (e.g. carbon-12 ions) and fission neutrons emitted from Cf-252, on the other hand, are referred to as high-LET radiation because they produce densely packed ionization/excitation events that gave high-LET values (>30 keV/μm). A high-LET radiation is more effective in cell killing than a low-LET radiation because the sub-cellular damage caused by the high-LET radiation is often too severe to be repaired correctly by the cell's inherent repair mechanisms.

The increased effectiveness of cell killing by high-LET radiations compared with low-LET radiations per unit absorbed dose is denoted by **r**elative **b**iological **e**ffectiveness (RBE). It is defined as the ratio of the dose of a standard photon beam (a 250 kVp X-ray beam or a Co-60 gamma-ray beam) to the dose of the test beam that is necessary to produce the same level of biological effect; i.e. it is an isoeffective dose ratio. In clonogenic survival assays, a survival fraction (SF) of 10% or 1% is often used as the "gold standard" for RBE estimations. As such, the RBE values vary with cell types. Clinically, the RBE values are estimated for both normal tissue reactions and tumor regression.

Other important considerations about the RBE values include:

1. The RBE of a beam varies with the absorbed dose, generally increases at low doses. As such, the clinical RBE has been assumed to increase as the dose per fraction is reduced [Joiner, 2002].

2. The RBE of a low-LET beam varies much more markedly with the absorbed dose rate than the RBE of a high-LET beam.
3. The RBE of a test beam depends on the reference photon beam. The original reference beam was 250 kVp X rays, but for the purposes of radiation oncology, the standard is now Co-60 gamma rays [Kal, 1986].

4.3 RBE of Cf-252 brachytherapy sources

The RBE of the total (i.e. neutron + gamma ray) emission of a Cf-252 brachytherapy source has been obtained via both in-vitro and in-vivo experiments [Kal, 1986]. In these experiments, one irradiate the sample object (in vitro or in vivo) separately with a ^{60}Co gamma-ray source and a Cf-252 source and then compare the doses of the two sources that are needed to achieve the same biological effect. The following formula was used to determine the RBE for the total (i.e. neutron + gamma ray) emission of a Cf-252 source:

$$(RBE_{n\gamma})_{Cf} = \frac{(D_\gamma)_{Co}}{(D_{n\gamma})_{Cf}} \tag{1}$$

where $(D_\gamma)_{Co}$ and $(D_{n\gamma})_{Cf}$ correspond to the gamma-ray dose from the Co-60 source and the mixed neutron and gamma-ray dose from the Cf-252 source, respectively. The total biological effect of a Cf-252 source can also be expressed as the sum of the neutron effect and the gamma-ray effect. That is,

$$(RBE_{n\gamma})_{Cf}(D_{n\gamma})_{Cf} = RBE_n(D_n)_{Cf} + (D_\gamma)_{Cf} \tag{2}$$

where RBE_n is the neutron RBE, $\left(D_{n\gamma}\right)_{Cf} = (D_n)_{Cf} + (D_\gamma)_{Cf}$, which is the total dose from the Cf-252 source, and $(D_n)_{Cf}$ and $(D_\gamma)_{Cf}$ are, respectively, the neutron dose and gamma-ray dose of the Cf-252 source. Equation (2) simply says that the total effect is the sum of the neutron effect and the gamma-ray effect. An inherent assumption of equation (2) is that the RBE of Cf-252 gamma rays is unity, i.e. the effect of Cf-252 gamma rays is identical to that of the Co-60 gamma rays in equation (1). The combination of equations (1) and (2) gives the following formula for neutron RBE:

$$RBE_n = \frac{RBE_{n\gamma}D_{n\gamma} - (D_\gamma)_{Cf}}{(D_n)_{Cf}} = \frac{(D_\gamma)_{Co} - (D_\gamma)_{Cf}}{(D_n)_{Cf}} \tag{3}$$

Since the biological effect of gamma rays increases significantly as dose rate increases, both RBE_n and $RBE_{n\gamma}$ vary according to dose rate of the Cf-252 source. Table 1 shows the broad range of values of RBE_n and $RBE_{n\gamma}$ for Cf-252 for cells in vitro [Kal, 1986]. The broad range of values of RBE_n and $RBE_{n\gamma}$ were also obtained for in-vivo normal tissue reactions and experimental tumors [Kal, 1986]. Tables 2 and 3 show these values for normal tissues and experimental tumors, respectively.

While the values of RBE_n and $RBE_{n\gamma}$ obtained from in-vitro and in-vivo studies vary greatly among cell/tissue types and they also depend greatly on dose rate, the clinical RBE values of a Cf-252 brachytherapy source for treating a specific tumor are much better defined. The clinical

RBE values are usually determined by comparing the maximum tolerable doses (MTD) of normal tissues under Cf-252 treatment to that under a conventional multifractional gamma-ray treatment. Tsuya et al initially considered the RBE_n to be 2.2-3.5; later changed to 5-7 [Tsuya et al., 1979]; Castro et al estimated a value of 6.5-7.5 [Castro et al., 1973]; Vallejo et al estimated the value to be 6.4 [Vallejo et al., 1977]; Paine et al used a value of 7.0 [Paine et al., 1976]. Vtyurin considered the RBE_n of mucous membrane, bladder and rectum to be 4.0-6.3 [Vtyurin, 1982]. For treatment of bulky tumors, Maruyama et al used 6.0 with the majority of tumors being of the cervix, vagina and uterus [Maruyama et al., 1983].

Biological system	Total dose rate (Gy h^{-1})	$RBE_{n\gamma}$	RBE_n
V79 cells	0.08		6.7
V79 cells	~0.135		3.9
V79 cells	0.18	3.5	5.9
CHO cells (5% SF)	0.15	5.1	6-7.1
HeLa cells	0.16	2.9	4
HeLa cells	~0.13		4.4
HeLa cells	0.20	2.2	2.7
T-1 cells	0.045-0.27		7.1
EMT-6 cells	0.05-0.27	5	3.0-5.3
CHL-F (prolif.)	0.013-0.052		8.7-20
CHL-F (1% SF)	0.05	5.9	9.2
CHL-F (1% SF)	0.02	9	14.3

Table 1. The values of RBE_n and $RBE_{n\gamma}$ for Cf-252 for cells in vitro.

Biological system	Total dose rate (Gy h^{-1})	$RBE_{n\gamma}$	RBE_n
Pig skin:			
Early reaction – moist desquamation	0.07-0.484		5.4-7.4
Early reaction – hair growth inhibition	0.12-0.13	5.7-7.2	8.5-10.8
Early fibrosis, vasculitis, muscle degeneration	0.04-0.08		3.8-4.6
Late fibrosis, vasculitis in rectum, cervix, bladder	0.04-0.08		6.6-7.0
Mouse:			
Early skin effects	3.6		1.54-2.32
Late skin effects	3.6		2.84-4.28
Jejunal crypt cells	0.44		4.9
$LD_{50/30}$	0.65		4.1
Bone marrow CFU	0.77		2.1-2.7
Bone marrow CFU	0.108		2.44
Bone marrow CFU	0.099		2.1
Testes	0.21	3.7	5.1

Table 2. The values of RBE_n and $RBE_{n\gamma}$ for Cf-252 for normal tissues.

Biological system	Total dose rate ($Gy\ h^{-1}$)	RBE_n
Mouse fibrosarcoma	0.13	6.3
Mouse EMT-6	0.78	4.7-5.4
Mouse leukemic cells	2.4; 2.8	2.7; 3.8
Mouse ascite cells	0.27; 0.0694	5.2; 5.6
Nude mouse HeLa tumors	1.86	11

Table 3. The RBE_n values of Cf-252 for experimental tumors.

In a Phase-I trial with Cf-252 used as the sole source of radiation in treating malignant brain tumors [Patchell et al., 1997], the MTD for the scalp and normal brain tissues are 900 cGy_n and 1200 cGy_n (i.e. cGy of neutron dose), respectively. One may obtain the neutron clinical RBE against a conventional **m**ulti**f**ractional (mf) gamma-ray treatment by replacing $(D_\gamma)_{Co}$ in equation (3) with $(D_\gamma)_{mf}$, the gamma-ray dose needed in a conventional gamma-ray treatment to achieve the same clinical effect. That is,

$$RBE_n = \frac{RBE_{n\gamma}D_{n\gamma} - (D_\gamma)_{Cf}}{(D_n)_{Cf}} = \frac{(D_\gamma)_{mf} - (D_\gamma)_{Cf}}{(D_n)_{Cf}} \tag{4}$$

If one assumes that the total MTD to the normal brain in the Cf-252 treatment includes 2/3 neutron dose (i.e. 1200 cGy) and 1/3 gamma-ray dose (i.e. 600 cGy) and that 8000 cGy is the MTD of normal brain tissue in a conventional multifractional gamma-ray treatment [Leibel & Sheline, 1991], then the RBE_n can be derived as follows:

$$RBE_n = \frac{(D_\gamma)_{Co} - (D_\gamma)_{Cf}}{(D_n)_{Cf}} = \frac{8000 - 600}{1200} = 6.2 \tag{5}$$

The above value is another confirmation that RBE_n of 6.0 is a good estimate to be used in treating a variety of tumors and that this value does not vary greatly among different types of normal tissues.

Lastly, radiobiological models have been developed to estimate the value of RBE for neutrons emitted from ^{252}Cf [Rivard et al., 2005; wang & Zhang, 2006; Wang et al., 2007]. These models are all based on the linear-quadratic (L-Q) formula for estimating cell survival fraction. One notable finding is that the cell survival fraction of a mixed irradiation of neutrons and gamma rays (e.g. the field produced by ^{252}Cf) is lower than the survival fraction of the cells irradiated with neutrons and gamma rays separately. The additional cell killing effect is attributed to the interactions between the sublethal lesions produced by the two different radiation types. It was suggested that the use of L-Q model to directly evaluate the isoeffect of a mixed neutron and gamma-ray irradiation is superior to the employment of RBE_n [Wang & Zhang, 2006]. This is because the L-Q model would eliminate the issue of what value of RBE_n should be considered as most appropriate.

4.4 Other biological justifications of neutron brachytherapy

The large RBE of a high-LET modality alone, however, does not make it superior to the low-LET gamma-ray therapy modalities. In fact, a therapeutic gain (TG) exists only if the RBE of tumor response to the high-LET beam is greater than the RBE of normal tissue response. As such, the advantage of a high-LET radiotherapy (e.g. NBT) is most pronounced in treating: (a) locally advanced tumors containing high proportions of hypoxic cells (i.e. cells being short of oxygen), (b) slowly growing tumors containing high proportions of cells that are in the radioresistant phases, and (c) tumors proliferating too fast to benefit from the conventional 6-week 30-fraction scheme. These include a large variety of tumor types: locally advanced prostate carcinoma, head and neck tumors, soft-tissue sarcoma, cancer of the cervix, colon, rectum, and esophagus, melanoma, and malignant glioma. Clinically, however, only a small number of patients in a few tumor types have so far shown clear benefit from neutrons: locally advanced salivary gland tumor (which represents a small fraction of head and neck tumors), locally advanced cervical cancer, and locally advanced prostate carcinoma, bone sarcomas and soft tissue sarcomas [Maruyama et al., 1997; Debus et al., 1998].

The therapeutic gain (TG) of a high-LET modality over the conventional low-LET photon modality in treating hypoxic tumors is often evaluated via the oxygen enhancement ratio (OER), defined as the radiation dose in hypoxic cells to that in aerated cells for the same biological effect:

$$\text{OER} = \frac{\text{Radiation dose in hypoxic cells}}{\text{Radiation dose in aerated cells}} \tag{6}$$

Most mammalian cell lines have an OER for cell killing of between 2.5 and 3 for photons [Kal, 1986]. It has been observed that the OER for Cf-252 irradiation is 1.4-1.6 [Kal, 1986]. If one takes the median values of the OERs, the potential therapeutic gain for ^{252}Cf NBT can be calculated as:

$$\text{TG} = \frac{(\text{OER})_{\text{photons}}}{(\text{OER})_{\text{Cf}-252}} = \frac{2.8}{1.5} = 1.87 \tag{7}$$

One of the obstacles that have hindered the progress of NBT was the late effect to normal tissues, which is believed to be more severe in all high-LET radiotherapy modalities than in the conventional x-ray therapy. That is, in the conventional x-ray therapy, much of the radiation damage to normal tissues is repaired during intermissions of the multi-fractional treatment scheme. In a high-LET radiotherapy, however, much of the damage to normal tissues is irreparable. This late-effect problem can be minimized by delivering superior dose distributions - i.e. large neutron doses to the tumor and small doses to the surrounding normal tissues. In fact, it has been shown that the late effect was significantly ameliorated among patients treated with the recently available multileaf collimators at several external beam fast neutron therapy (EBFNT) centers [Lindsley et al., 1998]. In addition, Maruyama concluded that among his first group patients ten years after being treated with the Cf-252 AT sources, serious complications were less than 5% [Maruyama, 1986]. The latest clinical study in China based on 696 patients with cervical cancer also shows that the incidence of late complications due to NBT is much lower than that for patients treated with the conventional brachytherapy using ^{192}Ir [Lei et al., 2011]. These evidences show that the late effect should no longer be an issue for patients treated with NBT.

In addition to the dose distribution issue, one other clarification needs to be made on the differences between the EBFNT and the NBT. Both modalities are based on neutrons. However, the average neutron energy in EBFNT (~30 MeV) is more than an order of magnitude greater than the average neutron energy in NBT (~2 MeV). Since the LET of neutrons increases as neutron energy decreases, the difference in neutron energy translates to a factor of 5 in the difference in LET. In other words, the LET of NBT is approximately 5 times that of the LET of the EBFNT. As such, it is more appropriate to label the EBFNT as an intermediate-LET modality rather than a high-LET modality. The NBT, on the other hand, is a true high-LET modality. This difference in LET, in turn, translates to the differences in the clinical values of RBE (2-3 for EBFNT and 6-7 for NBT) [Kal, 1986] and OER (1.7 for EBFNT and 1.5 for NBT) [Kal, 1986]. Consequently, in comparison with the EBFNT, the NBT is more effective especially in treating hypoxic tumors.

5. Clinical data on neutron brachytherapy

Clinical trials involving NBT have been conducted in six countries during 1969-1997 to test the safety and efficacy of various protocols in an attempt to improve the therapeutic control of bulky, radioresistant tumors [Maruyama et al, 1997]. The resultant published data contains more than five thousand cases of patients treated in the United States, Russia, Czechoslovakia, Lithuania, Japan and England. During this period of research, twelve different groups, comprising more than eighty-nine physicians, physicists, oncologists and radiation biologists, conducted over sixty trials. Patients who participated in these studies were followed from two months to fifteen years to determine the overall clinical outcomes. In 1999 China became the first country taking NBT to the commercial stage, and has since treated approximately 20,000 patients with majority of them having cervical cancers [Lei et al., 2011].

Many of the early trials involving the use of Cf-252 brachytherapy attempted to determine the most effective scheduling protocol. Each group examined various dose schedules, with Cf-252 most commonly being used as an "early" or "late" neutron boost in conjunction with conventional external beam therapy. The validity of Cf-252 as a single modality was also researched. As a general consensus, all groups found Cf-252 to be extremely effective in causing rapid tumor regression in bulky, localized radioresistant tumors with large numbers of hypoxic cells. The effective regression of tumor size caused by Cf-252 results from the high-LET fission neutrons released during treatment. The sites in the studies included malignant glioma, gynecological tumors, cancers of the head, neck and oral cavity, gliomas. Each research group used sources of varying sizes and strengths depending on the location and year of source fabrication. Studies conducted in the United States and Japan used Cf-252 sources fabricated at the Savannah River Site (SRS) or Oak Ridge National Laboratories (ORNL). Studies performed in Russia, Czechoslovakia, and Lithuania used sources originating from Dmitrovgrad City, Russia.

5.1 Cervical treatment evaluation

Since the early 1970's, cervical cancer has been recognized as an ideal site for use of Cf-252. Because the ORNL "AT" source size is similar to that of existing photon sources such as Ra-

226 or Cs-137, it became a matter of substitution rather than whole design changes when it came time to use Cf-252.

Sixteen different groups in five different countries have carried out cervical cancer treatments with Cf-252 brachytherapy [Maruyama et al., 1997]. Various dose schemes, including a majority of trials that investigated the application of Cf-252 before ("early") and after ("late") external beam radiotherapy, were utilized in order to determine overall treatment efficacy. Other trials also examined the combination of Cf-252 and Co-60 or Ra-226. Patients were typically followed a minimum of two years with one trial following patients up to twelve years. Outcomes generally are cited in one of three groups, including tumor regression (TR), local tumor control (LTC) and survival rate (SR) with complications being sited when applicable. Survival rates for the use of Cf-252 throughout the various trials brought to light the fact that earlier stages of the disease had a higher degree of treatment success as compared to later stages. A definite advantage for the use of Cf-252 was also displayed over conventional gamma and external beam therapy in a majority of cases. Complications that have been observed through the course of these studies include, but are not limited to ulcerative cystitis, epithelitis, fistulas, rectal bleeding, rectitis, enterocolitis and hematuria.

A representative example of the trials that were conducted includes a study by a Russian group led by Marjina that treated 1,055 patients with stage I through III endometrial cancers from 1983-1996 [Marjina et al., 1997]. The study compared the efficacy of "early" and "late" application of Cf-252 in conjunction with external beam therapy and as a single treatment modality. The patients entered in the trial were followed for twelve years, with final results indicating that the LTC for "early" Cf-252 application was 87.2%, compared to 90.1% for "late" Cf-252 application and 79.4% for use of Cf-252 as a single modality. Five year survival rates were also determined for the three dose schemes, in which "early" versus "late" versus single application resulted in 86.5%, 86.3% and 56.1% respectively. Ten year survival rates for stage II versus stage III resulted in 71.6% and 66.6% respectively.

Another study by a Czechoslovakian group led by Tacev from 1986-1992 treated 430 patients with stage II and III cervical cancers examining the cumulative effects of Cf-252, Ra-226 and external beam therapy [Tacev et al., 1997]. Three dose schemes were utilized in which group: a) received 16 Gy of neutrons, 40 Gy of Ra-226 and 40 Gy of external beam therapy, group b) received 40 Gy of neutron, 16 Gy of Ra-226 and 40 Gy of external beam therapy and group c) simply received a conventional treatment regimen comprising 56 Gy of Ra-226 and 40 Gy of external beam therapy. Patients were followed for five years, with survival rates for groups a), b) and c) with stage II resulting in 89.8%, 83.7% and 72% respectively, and stage III resulting in 67.7%, 61.8% and 45% respectively.

An American group led by Maruyama at the University of Kentucky conducted the final representative study in which 218 patients with stage I, II and III cancer were treated by an "early" and "late" dose scheme in conjunction with external beam therapy from 1976-1983 [Maruyama et al., 1991]. Patients enrolled in the study were followed for up to ten years. Five year survival rates for stage I, II and III were 87%, 62% and 33% respectively, while ten year survival rates were 82%, 61% and 25%.

However, when the Cf-252 data from Lithuania and Japan are reviewed, one can see that the local control and survival rates are similar but complication rates are higher [Shpiklov et al.,

1991; Yamashita et al., 1991]. The apparent reason lies in the dose rates that were utilized in these centers. They were about 40% higher and/or were given on a more frequent schedule. The lower dosing schedules put forth early on at the University of Kentucky by Dr. Maruyama were not only efficacious, but had the highest level of safety as it related to complications.

The most important study is a randomized Phase III trial conducted by Tacev and colleagues from the Czech Republic [Tacev et al., 2003]. In that trial over 100 women with St IIb, IIIa, or IIIb received identical treatment with the exception that the initial brachytherapy was randomized between Cf-252 (6Gy-eq = 40 Gy) or CS/Ra. In this trial the 5 year local control ($p < 0.00009$), 5 year overall survival ($p < 0.001$) favored those receiving Cf-252. In addition there was no difference in complication rates between the two groups of patients. Clearly, Cf-252 had superior outcomes and yet had similar complication rates as Cs-137/Ra-226.

The above cited trials were conducted without use of chemotherapy, which has become a standard of care in the US. However, in 1993 the University of Kentucky group published on a cervical cancer Phase II trial in which both CDDP (50 mg/m^2) and 5FU (1000mg/m^2) were delivered with twice-daily radiation (120 cGy BID) and Cf-252 brachytherapy [Maruyama et al., 1993]. The equivalent doses to Point A (85 Gy) and Point B (60 Gy) are what is accepted for traditional photon brachytherapy source treatment. In that trial only 1 patient had a grade III/IV event, well below other institutional reports using similar techniques with photon irradiation alone. Again, there is strong evidence of efficacy and safety from the proper use of Cf-252 in the treatment of cervical carcinoma.

From February 1999 to December 2007, 696 patients with cervical cancer (Stages IB to IIIB) in China were treated with NBT in combination with external-beam therapy [Lei et al, 2011]. The NBT was delivered at 7-12 Gy per insertion per week, with a total dose of 29-45 Gy to reference point A in three to five insertions. The RBE value used was between 2 and 3. The whole pelvic cavity was treated with 8-MV X-ray external irradiation at 2 Gy per fraction, four times per week. After 16-38 Gy of external irradiation, the center of the whole pelvic field was blocked with a 4-cm-wide lead shield, with a total external irradiation dose of 44-56 Gy. The total treatment course was 5 to 6 weeks. The overall survival rate at 3 and 5 years for all patients was 76.0% and 64.9%, respectively. Disease-free- 3- and 5-year survival rates were 71.2% and 58.4%, respectively. Late complications included vaginal contracture and adhesion, radiation proctitis, radiation cystitis, and inflammatory bowel, which accounted for 5.8%, 7.1%, 6.2% and 4.9%, respectively. These results compare favorably to the results of patients treated with the conventional high-dose-rate photon brachytherapy using ^{192}Ir.

5.2 Head and neck treatment evaluation

The head and neck region covers multiple cancer treatment sites including esophageal, tongue, floor of mouth, lip, skin/soft tissue, buccal mucosa, oral cavity and oropharynx. Twelve different groups in four countries have carried out head and neck cancer treatments with NBT. A majority of dose schemes used in these studies combined NBT with conventional external beam radiotherapy, with a handful of trials comparing Cf-252 and Co-60. Patients were typically followed a minimum of two years with a few trials following patients up to ten years. Outcomes generally are cited in one of three groups, including tumor regression, local tumor control, and survival rate with complications being sited when applicable. The following studies detail the dose schemes, the year of the study, and the overall results obtained during the follow-up periods for select trials that are a representative sample of the head and neck cancer treatment group.

The first clinical study was conducted from 1973-1986 by a Russian group led by Vtyurin that treated 488 patients with a variety of primary and recurrent stage T1-T4 head and neck cancers, including tongue, oral cavity, buccal mucosa, oropharynx, lip and skin/soft tissue [Vtyurin, et al., 1997]. The dose scheme utilized in the trial examined the use of Californium-252 before ("early") and after ("late") external beam therapy and also as a single treatment modality. Patients enrolled in the study were followed for ten years with three and ten year survival results for the various treatment sites being as follows: Primary Tongue=57.9%, 36.5%; Recurrent Tongue=47.9%, 29.1%; Primary Oral Cavity=54.8%, 32.3%; Recurrent Oral Cavity=41.1%, 17.6%; Primary Buccal Mucosa=56.2%, 37.5%; Recurrent Buccal Mucosa=No data available; Primary Oropharynx= 44.4%, 27.7%; Recurrent Oropharynx=46.2%, 30.7%; Primary Lip=92.3%, 82%; Recurrent Lip=59.2%, 25.9%; Primary Skin/Soft Tissue=77.8%, 61.1%; Recurrent Skin/Soft Tissue=86.3%, 22.7%.

A Lithuanian group led by Shpiklov conducted another trial from 1987-1989, in which 43 patients with stage I tongue and oral cavity cancers were treated using a mixed therapy regimen of Cf-252 and fractionated external photon beam [Shpiklov et al., 1991] . The results of the study showed a tumor regression rate of 63% for the tongue cancer and 83% for the oral cavity cancer.

5.3 Malignant glioma treatment evaluation

Chin conducted the first clinical study of using Cf-252 to treat malignant glioma at the University of Kentucky from 1980-1990 [Chin et al., 1991]. In this study, 85 patients were treated solely using single and multiple Cf-252 tubes and also in conjunction with external beam therapy. The overall median survival time with Cf-252 being used as a single modality was 12 months. The results obtained from the single Cf-252 tube in combination with external beam therapy at the eight month point had a survival rate of 87%, but rapidly decreased after that point. The multiple Cf-252 tubes in combination with external beam therapy however, were found to have a one year survival rate of 73%, with only a gradual decrease after that point indicating a definite advantage over single tube use.

Patchell et al continued Chin's work at Kentucky with an open ended Phase I trial (a dose searching study), to test the feasibility of NBT as the sole treatment modality, and to determine the maximum tolerable dose for the treatment of malignant gliomas [Patchell et. al., 1997]. The study was an open-ended dose escalation study. Radiotherapy was delivered using Cf-252 implants as the sole source of radiation with a dose escalation protocol to determine the maximum effective tolerable dose. A starting dose of 900 cGy_n (i.e. 900 cGy of neutron, which is equivalent to about 5400 cGy of gamma-ray) was known to be well tolerated and was chosen as the starting point. Thirty-three (33) patients with histologically confirmed astrocytomas or glioblastoma multiforme were entered into the study. All patients had undergone previous debulking surgery. Three patients with newly diagnosed malignant glioma were entered at each dose step, and the number increased to six patients in dose steps at which necrosis of brain occurred. Ten patients developed scalp necrosis associated with scalp doses above 900 cGy_n. The protocol was revised to limit the maximum scalp dose to less than 900 cGy_n. After this reduction, no additional patients developed scalp necrosis. The study ended when two patients at the 1300 cGy_n dose step developed radiation necrosis of the brain. In all, three patients developed brain necrosis outside the treatment volume (more than 2 cm beyond the lesion as determined by enhanced MRI scan), one at a dose of 1200 ncGy and two at a dose of 1300 ncGy. In every instance, the brain necrosis was confined to the area adjacent to the treatment volume, and the extensive necrosis seen after external beam neutron therapy was not present.

The clinical investigators concluded:

1. Neutron brachytherapy using Cf-252 as the sole source of radiation is a feasible treatment for malignant glioma;
2. The scalp tolerates less neutron radiation than the brain; and
3. The most tolerated dose (MTD) (and the recommended dose for a Phase II trial) of interstitial neutron brachytherapy is 1200 cGy_n.

Use of this starting point avoided ineffective doses while avoiding radiation necrosis in the first dose step. After a given dose step had been completed, the next escalation step was not started until the first patient treated in the previous dose step had completed neutron therapy and had been observed for at least 12 weeks. Each subsequent dose was increased by 100 cGy_n to 1300 cGy_n. For study purposes, the patients were followed indefinitely until they developed radiation necrosis of brain or died. Three to ten implant tubes (median, 6) with three to fourteen (median, 10) Cf-252 sources per implant were used. The median implant duration was 29.4 hours, with a range of ten to sixty-five hours. Early in the study, scalp necrosis was identified as an important complication. The median survival time was 10.9 months with five patients exhibiting no tumor recurrence at the time of death. Of these five patients however, three had radiation necrosis of the brain and two had infections associated with radiation necrosis of the scalp. The frequency of scalp necrosis throughout the study was noted as being higher than anticipated, with two possible explanations for the cause. The first explanation postulated, was the high fat content in the scalp region, which would have provided a high density of hydrogen atoms resulting in a higher neutron dose than other soft tissues. The second explanation was the inadvertent extension of the neutron implants beyond the margin of the brain to the level of scalp, which would have accounted for the higher dose observed and the resulting scalp necrosis. The early complication of scalp necrosis was subsequently controlled with careful treatment planning, and scalp doses were kept at less than 900 cGy_n.

5.4 Rectal treatment evaluation

Three different groups from two countries have carried out rectal cancer treatments with NBT on 64 patients at various times from the 1980's to 1990's [Burneckis et al., 1997; Sidorchenkov et al., 1997]. Multiple dose schemes were utilized depending upon the overall treatment strategy being used. Californium-252 was used in a mixed therapy regimen with external beam radiotherapy, as well as preoperatively as a single modality to enhance tumor regression. Results from the studies indicated a tumor regression rate of 83% in one study, with only anticipated complications arising in all three trials such as mild bleeding and minimal damage of proctal mucosa.

5.5 Hallmarks of performance

The clinical studies conducted to-date have consistently produced evidence to substantiate the following performance characteristics as hallmarks of NBT.

5.5.1 Single inactivation of cells

The neutrons of Cf-252 have very high LET. This means that a single event can inactivate the target site in the cell and lead to one-hit inactivation of cells. Clinical observations indicate high sensitivity and quick response of tumors to Cf-252 neutrons.

5.5.2 Absence of dose rate effect

Clinical studies show no dose rate effects. These results were expected based upon the one-hit survival curves of cells irradiated with Cf-252. The absence of dose rate effects allows Cf-252 therapy to be planned for a period of approximately 4 to 8 hours rather than the standard protracted and lengthy 48 to 144 hour schedules (at approximately 40 to 50 cGy per hour) used for photon radiation. Clinical observations show that there is little effect of varying dose rates in tumor response by Cf-252 studies to date.

5.5.3 Absence of sub-lethal (SLD) repair

One-hit survival curves of cells irradiated with Cf-252 mean little or no SLD repair occurs. No SLD repair is advantageous where neutron effects in the tumor result from Cf-252 sources implanted directly into the tumor. By avoiding or minimizing neutron radiation of adjacent sensitive organs, few complications were observed.

5.5.4 Late effect and leukemogenesis

Clinical trials have shown only about a 5% normal tissue complications have been observed after NBT with follow-up through 15 years. Further, there has been no occurrence of late leukemias, aplastic anemias, myelomas, or lymphomas observed. Secondary tumors in the lung or bowel are consistent with historical photon brachytherapy experience and do not present as an additional risk.

5.6 Summary of clinical results

Nearly 40 years of published clinical research of more than 25,000 patients treated in seven countries has demonstrated the viability of using Cf-252 NBT to treat cancerous tumors. Californium-252 NBT was found to be highly effective in causing rapid tumor regression in bulky, localized radioresistant tumors with large numbers of hypoxic cells. The effective regression of tumor size caused by Cf-252 is due to the high–LET nature of the spontaneous fission neutrons released during treatment. The absence of any new or serious side effects (when compared to photon brachytherapy) provides further argument that NBT is a viable alternative to, and an equal in performance to, photon brachytherapy. The five-year survival rates are equal to, and in some cases greater than, those seen with photon brachytherapy. It is reasonable to expect that the newly developed high-intensity miniature Cf-252 sources will improve dose delivery (over the large AT sources) and thus further improve the clinical outcomes of NBT.

6. Californium-252 propertities and dose determination

6.1 Decay and emissions

Californium-252 has a half-life of 2.645 years. The majority, 96.9%, of ^{252}Cf decay is through alpha decay, but due to the short range these alpha particles do not escape the source capsule. A small, 3.1%, of ^{252}Cf decay is through spontaneous fission, and each fission produces two or three fission fragments as well as an average of 3.77 neutrons [Wierzbicki et al, 1997]. One μg of ^{252}Cf (0.536 mCi) emits 2.31 x 10^6 neutrons sec^{-1} (approximately 2 percent overall uncertainty) [ICRU, 1977]. These fission neutrons have an energy spectrum that is often modeled as either a Maxwillian or a Watt fission spectrum, which peaks at 0.7 MeV and falls off rapidly at both higher and lower energies (see Fig. 3). The mean neutron energies inferred from reported measurements are between 2.13 and 2.15 MeV [Grundl & Eisenhauer, 1975; Walsh, 1991].

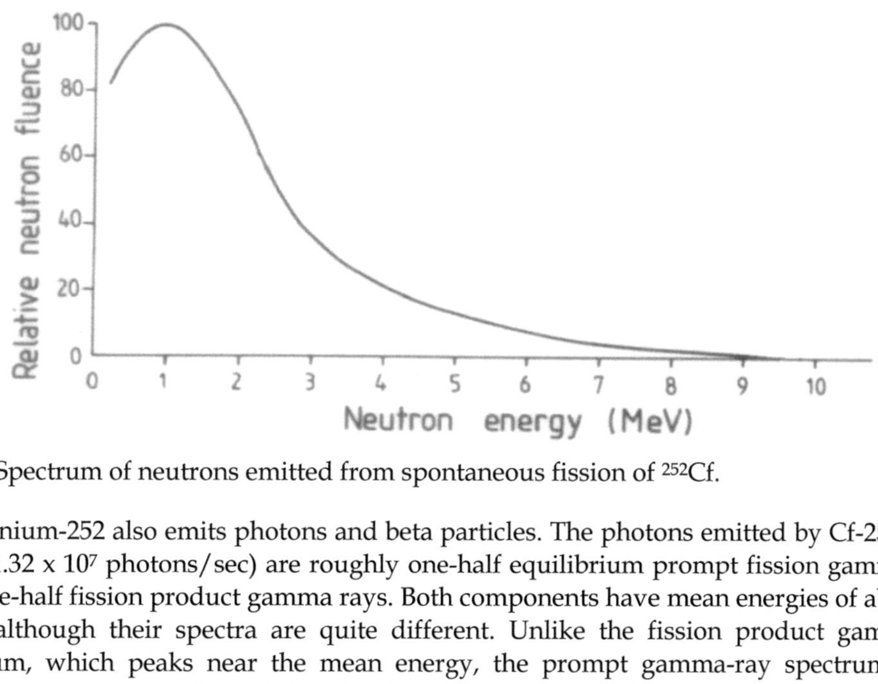

Fig. 3. Spectrum of neutrons emitted from spontaneous fission of ^{252}Cf.

Californium-252 also emits photons and beta particles. The photons emitted by Cf-252 (1 μg emits 1.32 x 10^7 photons/sec) are roughly one-half equilibrium prompt fission gamma rays and one-half fission product gamma rays. Both components have mean energies of about 0.8 MeV, although their spectra are quite different. Unlike the fission product gamma-ray spectrum, which peaks near the mean energy, the prompt gamma-ray spectrum has a significant component above 3 MeV and increases exponentially with decreasing energy. Table 4 shows the gamma-ray energy spectra. The gamma-ray associated with alpha-particle decay is negligible (<0.1%). Because the fission products gradually builds up in a sealed Cf-252 source, one might expect that the absorbed dose of gamma rays relative that of neutrons emitted from the source would change with time. This, however, is not the case. The absorbed dose ratio was found to be a constant within a standard deviation of ±5% over a period of about three years [ICRU, 1977]. Given the 2.64 year half-life, any build-up of fission products is, therefore, unimportant.

In general, the beta particles (emitted from the fission products) and the associated bremsstrahlung x-rays do not penetrate the source capsule, and therefore, do not contribute to the external dose. However, for the new miniature sources having capsule wall thickness of 0.2 mm, the bremsstrahlung x-rays do penetrate the source capsule and contribute significantly to the dose in the immediately vicinity of the source. Detailed discussion on the bremsstrahlung x-ray dose component for the new miniature sources is provided in Section 6.3.

Energy (MeV)	Gamma rays emitted from prompt fissions	Gamma rays emitted from fission products under equilibrium	Total
0.0-0.5	3.3×10^6	1.3×10^6	4.6×10^6
0.5-1.0	1.7×10^6	4.0×10^6	5.7×10^6
1.0-1.5	7.7×10^5	1.9×10^5	1.7×10^6
1.5-2.0	4.2×10^5	3.5×10^5	7.7×10^5
2.0-2.5	2.2×10^5		2.2×10^5
2.5-3.0	1.1×10^5		1.1×10^5
3.0-3.5	5.6×10^4		5.6×10^4
3.5-4.0	3.0×10^4		3.0×10^4
4.0-4.5	1.7×10^4		1.7×10^4
4.5-5.0	8.2×10^3		8.2×10^3
5.0-5.5	4.9×10^3		4.9×10^3
5.5-6.0	1.8×10^3		1.8×10^3
6.0-6.5	1.0×10^3		1.0×10^3
		Total = 1.322×10^7 photons μg^{-1} sec^{-1}	

Table 4. Spectra of gamma rays emitted from spontaneous fission and the fission products of Cf-252.

6.2 Dosimetry protocol

The absorbed dose of neutrons in tissue is mainly deposited by the recoil hydrogen nuclei (i.e. protons) from elastic scattering interactions. In average, a neutron transfers approximately half of its kinetic energy to a recoil proton in an elastic scattering interaction. After losing all its kinetic energy, the neutron is soon captured either by hydrogen via the $^1H(n,\gamma)^2H$ reaction or by nitrogen via $^{14}N(n,p)^{14}C$ reaction. As mentioned in Section 4, because the recoil protons deposit (or transfer) all their energies in tissue by producing short track (<100 μm) of densely packed ionization events, the absorbed dose of neutrons is referred to as being high-LET.

The absorbed dose of gamma rays in tissue is mainly deposited by the recoil electrons from Compton Scattering interactions. In average, a gamma photon transfers approximately one third of its energy to a recoil electron in a Compton Scattering interaction. After losing much of its energy in a few Compton Scattering interactions, the gamma photon is soon captured via a photoelectric absorption. As mentioned in Section 4, because the recoil electrons deposit their kinetic energies in tissue by producing long tracks of sparsely distributed ionization events, the absorbed dose of gamma rays in tissue is referred to as being low-LET. The rule of thumb is that roughly one-third of the absorbed radiation dose (Gy) in tissue near a Cf-252 source is due to low-LET gamma rays and two thirds is due to high-LET neutrons.

Since an effective brachytherapy treatment relies highly on accurate prescription of absorbed dose in and around a tumor volume, the objective of the NBT dosimetry protocol is to establish a method (or methods) that accurately specifies in detail the three-dimensional spatial distributions of absorbed dose rate in tissue-like medium (e.g. water) surrounding the source. The three-dimensional dose (or dose rate) distributions may be measured and calculated in a number of ways. The following subsections describe two different methods that were both used to obtain the detailed neutron and gamma-ray dose distributions surrounding a Cf-252 NBT source. Because the two methods are fundamentally different, their results should be complementary to each other. As such, the degree of agreement between the results may serve as a measure of the accuracy (or uncertainty) of the results.

6.2.1 The combined Monte Carlo and source strength measurement method

This method combines both the computational and the measurement results to obtain the dose distributions in water near a Cf-252 source. It first uses the Monte Carlo radiation transport code, MCNP-5 [J.F. Briesmeister, 2000], to calculate the normalized neutron and gamma-ray dose distributions in water (or tissue) nearby the source. It then uses a measurement method to obtain the neutron source strength (neutrons sec^{-1}) of the source. The absolute dose (or dose rate) distributions are then obtained by the multiplication of the two results.

A Monte Carlo radiation transport code is a computer-based method. It uses the known data associated with the radiation particles (e.g. neutrons, gamma rays, and electrons) and their interaction probabilities to simulate the random-walk behaviors of the radiation particles in the media. In a Monte Carlo code such as MCNP, the random-walk "history" of each particle is truly followed from birth to death. The information associated with a particle "history" at any moment throughout its lifetime includes particle type, energy, position, and direction. The absorbed doses at various locations (i.e. the dose distributions) in the media are obtained by tallying the statistics accumulated from a large number of particle histories. Based on today's computer speed (with a typical PC), it only takes a few minutes to run through millions of particle histories and to obtain results with statistical uncertainties of less than 1%. In addition, the MCNP code is the most widely used Monte Carlo code in the world for performing neutron, gamma-ray, and electron transport calculations, and therefore, both the algorithm and the data used in the code has been thoroughly verified and validated for its accuracy. Figure 4 shows the MCNP results of the neutron isodose contours normalized to 1 cm traverse the new miniature NBT source. The MCNP results have been compared with the previously obtained results and achieved good agreement [Rivard, 1999].

Since the MCNP results are normalized to a unit quantity of Cf-252 (e.g. 1 μg), to obtain the absolute dose rates one needs to multiply the MCNP results with the source strength (i.e. neutrons sec^{-1}) that can only be determined by measurement. The measurement method used at ORNL was based on three U-235 fission chambers surrounding the source. Each neutron emitted from the source has a fixed probability to be absorbed by a U-235 atom in one of the fission chambers, and thus produces a count. The measured neutron count rate of the source was then compared with the count rate obtained from a source of which the strength has been previously determined at the National Institute of Standard Technology

(NIST). The strength of the source was then derived from the measured count-rate ratio of the two sources.

The strength of a Cf-252 source determined at NIST is based on the Manganese Sulfate ($MnSO_4$) Bath Method [McGarry & Boswell, 1988]. In this method, the source is placed at the center of a large spherical tank, 1.27 m in diameter, containing $MnSO_4$ solution with a density of 1.37 kg liter^{-1}, until equilibrium of the Mn-56 activity is reached. This takes about 25 hours. The neutron emission rate of the source is determined by comparing the measured event rate of the gamma rays from the Mn-56 to that of the NIST standard source, NBS-I. The measured neutron emission rate (in neutrons sec^{-1}) is then converted to the Cf-252 content (in μg) using the conversion factor, 2.134 x 10^6 neutrons sec^{-1} μg^{-1}. The source strength obtained from this method has an uncertainty of 1.2%.

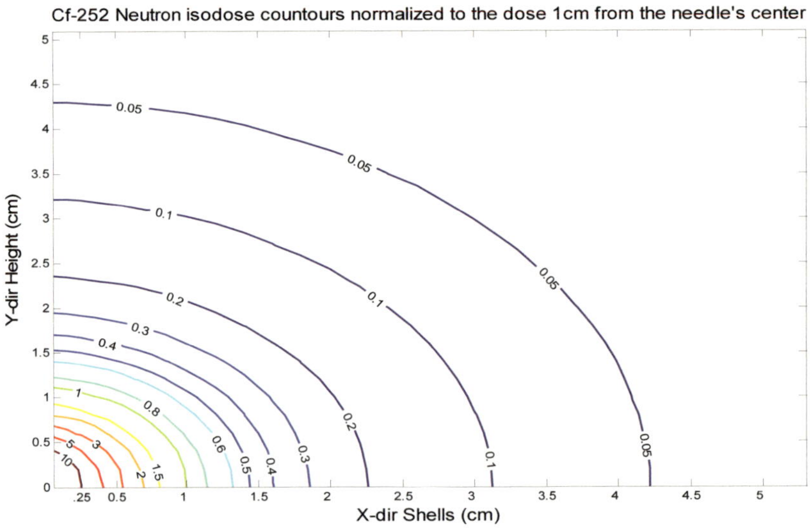

Fig. 4. Neutron isodose contours normalized as factors of the dose 1 cm traverse the source.

6.2.2 The two-ion chamber method

The most widely used method directly measuring both neutron and gamma-ray dose distributions in a mixed neutron and gamma field is the "two-ion chamber method". In this method, two miniature ion chambers having different neutron and gamma-ray responses are used to obtain neutron and gamma-ray doses separately [ICRU, 1977]. The two ion chambers commonly used are the T1 and M1 ion chambers made by Standard Imaging, Inc. Both chambers have the same geometry and a 0.056 cm^3 collecting volume. Of the two, T1 has high sensitivities to both neutrons and gamma rays, whereas M1 has a high sensitivity to gamma rays but a very low sensitivity to neutrons. The response of each ion chamber is related to neutron and gamma-ray doses D_n and D_γ by the following equations:

$$R'_T = k_T D_n + h_T D_\gamma \tag{8}$$

$$R'_M = k_M D_n + h_M D_\gamma \tag{9}$$

where the subscripts T and M refer to the ion chambers T1 and M1, respectively. Accordingly, R'_T and R'_M are the responses (i.e. the electric charge) of T1 and M1, respectively. The coefficients, k_T, k_M, h_T, and h_M, are the corresponding neutron sensitivities and gamma-ray sensitivities for T1 and M1. The values of h_T and h_M can be experimentally determined using a NIST-traceable Co-60 gamma-ray source, and the common unit used is nC/cGy. The values of k_T and k_M can be obtained by using the relative response curves available from ICRU-45 [ICRU, 1989]. One may then obtain the absorbed doses, D_n and D_γ, by solving equations (8) and (9) as:

$$D_n = \frac{h_M R'_T - h_T R'_M}{h_M k_T - h_T k_M} \tag{10}$$

$$D_\gamma = \frac{k_T R'_M - k_M R'_T}{h_M k_T - h_T k_M} \tag{11}$$

To obtain the dose distribution data for various positions near a Cf-252 brachytherapy source, one must conduct an in-phantom experiment with the two ion chambers, T1 and M1. In this experiment, the source is placed via a catheter tube located at the center of a Lucite-walled water phantom. The two ion chambers are placed under water at various locations and distances from the source to collect the electric charges. The electric charge data are then converted to the corresponding neutron and gamma-ray doses at various positions using equations (10) and (11). The spatial resolution of dose distribution is limited by the size of the ion chambers collecting volume, which is about 4 mm. This resolution is adequate for far away positions, but too poor for the nearby positions (< 2 cm from the source center point) due to the sharp dose gradients. For the nearby positions, one may use thermoluminescent dosimeters (TLD) or silicon diodes to measure the gamma-ray dose [ICRU, 1977].

6.3 Increased gamma dose in the new miniature source

A recent study of neutron and gamma dose profiles in water near a new miniature ^{252}Cf brachytherapy source reported discrepancies between the measured dose profiles and the dose profiles obtained computationally [Wang & Kelm, 2009]. The measured gamma dose rates near the source were found to be slightly greater than the neutron dose rates, contradicting the well established neutron-to-gamma dose ratio of approximately 2:1 at locations near a ^{252}Cf brachytherapy source of the old design [Rivard, 1999]. Specifically, the MCNP-predicted gamma dose rate is a factor of two lower than the measured gamma dose rate at the distance of 1 cm, and the differences between the two results gradually diminish at distances farther away from the source. These discrepancies were investigated and successfully explained by the effect of bremsstrahlung X-rays produced by the beta particles emitted from the fission products contained in the source. The well established neutron-to-gamma dose ratio of 2:1 works well for the old and large sources, but does not work for the new miniature sources. For the AT tube, the source capsule is thick enough that few bremsstrahlung photons can escape the capsule and contribute to the dose outside. For the new miniature source, however, because the capsule wall is so thin that many bremsstrahlung photons do escape the capsule wall and contribute significantly to the dose outside. Indeed, when one includes bremsstrahlung into the MCNP calculations, the gamma dose rates obtained from calculations agree well with the measured results [Fortune et al., 2011].

7. Dose prescription and treatment planning

Because the Cf-252 emission includes both neutrons and gamma rays and because neutrons are many times more effective than gamma rays (per unit dose) in cell killing, the quantity used for dose prescription is based on that of equation (12). The quantity is often referred to as gray equivalent, or Gy_{eq}. That is,

$$Gy_{eq} = RBE_n D_n + D_\gamma \tag{12}$$

where D_n and D_γ are respectively the neutron dose and gamma-ray dose, and RBE_n is the relative biological effectiveness of neutrons with respect to gamma rays. In other words, Gy_{eq} is the equivalent amount of gamma-ray dose to achieve the same biological effect. As discussed in Section 4.3, clinical experience shows that RBE_n of 6 is a good choice for setting the maximum tolerable dose of normal tissues for many tumor types.

The method for dose delivery and treatment planning method can be directly borrowed from that of the existing Ir-192 high-dose-rate (HDR) remote afterloading systems with little modification. In other words, the treatment planning softwares such as BrachyVision of Varian and PLATO of Nucletron are directly applicable to NBT. The only modification needs to be made is the dose distribution data. That is, the gamma-ray dose distribution surrounding the HDR ^{192}Ir source will be replaced by the Gy_{eq} distribution surrounding the Cf-252 source. It should be noted, however, that the Gy_{eq} distribution are pre-calculated and pre-calibrated in water (see Section 6). The actual dose (in Gy_{eq}) to a tissue during a treatment will differ slightly from the dose to water. This is because neutrons strongly interact with hydrogen nuclei and because the hydrogen contents of tissues (especially fatty tissue) differ slightly from that of water. Table 5 shows the neutron dose at 1 cm from an AT source for various tissues including water [Rivard, 1999]. As shown, the neutron dose of most tissues is 6-8% less than that of water except that of fat, of which the neutron dose is 6% greater than that of water. The neutron dose to the bone is much less because the water (i.e. hydrogen) content is much less than that in other tissues.

Tissue Material	Neutron dose ($cGy\ hr^{-1}\ \mu g^{-1}$)	Dose factor normalized to water
Water	2.064	1.000
Muscle	1.907	0.924
Brain	2.009	0.973
Skin	1.91	0.925
Fat	2.183	1.058
Blood	1.928	0.934
Lung	1.943	0.941
Bone	0.809	0.392

Table 5. Neutron dose at 1 cm from an AT source for various tissues including water [Rivard, 1999].

8. Boron-Enhanced Neutron Brachytherapy

Boron-enhanced neutron brachytherapy (BENBT) is a combination of neutron brachytherapy (NBT) and boron neutron capture therapy (BNCT). In a BNCT, a boron compound (a neutron capture agent) is first administered into the tumor of a patient. The tumor volume is then irradiated with slow (or thermal) neutrons. Each neutron capture reaction, $^{10}B(n,\alpha)^{7}Li$, releases an average of 2.4 MeV via short-ranged heavy ions, and therefore, is highly effective in killing the cell within which the reaction takes place. BNCT by itself, however, has not yet achieved notable clinical success mainly because it demands a perfect boron compound that can enter tumor cells with a high degree of specificity as well as a high concentration. The BENBT reduces that demand in that it is based on the already successful NBT with an additional boost of BNCT. No additional neutron irradiation is needed. The boost of BNCT comes from the same neutrons emitted from the ^{252}Cf brachytherapy source that are thermalized within the tumor volume.

In an experimental study [Wanwilairat et al., 2000] that employed a large water phantom (53 cm x 56 cm x 40 cm) with a 100-μg ^{252}Cf source (5.5 mm in active length and 3.3 mm in active diameter) placed at the center of the phantom, the results how that the relative ^{10}B enhanced dose increases linearly as a function of distance and reaches its maximum at 9.5 cm. If one assumes a ^{10}B concentration of 50 ppm, this maximum dose enhancement is translated to an increase of 28% of total dose. In another computational study [Rivard & Zamenhof, 2004], assuming a 15-cm-diameter brain phantom with a ^{10}B loading of 30 ppm and a tumor:healthy tissue ^{10}B ratio of 3:1, the results show that the biologically weighted dose enhancements are 0.6%, 6.5%, and 12.7% at 1, 3, and 5 cm (from the ^{252}Cf source), respectively. If one assumes that brain tumor recurrence is observed at 5 cm, then a 12.7% biologically weighted dose enhancement may be considered significant from clinical perspective. If new ^{10}B compounds were developed which could deliver greater tumor loading and tumor specificity, then BENBT would certainly become more clinically attractive.

9. Acknowledgement

The author would like to acknowledge Isotron Inc. for funding the study on dose characterization of the new miniature Cf-252 brachytherapy source.

10. References

Anderson, L. (1973). Status of Dosimetry for ^{252}Cf Medical Neutron Sources," *Phys. Med. Biol.* Vol.18, pp. 779-799.

Briesmeister, J. (2000). MCNP - A General Monte Carlo N-Particle Transport Code System, Version 5, LA-12625-M.

Burneckis, et al. (1997). Preoperative Radiotherapy in the Treatment of Rectal Carcinoma, In: *Californium-252 Isotope for 21st Century Radiotherapy*, Wierzbicki, J. (Ed.), pp. 199-201, Kluwer Academic Publishers, ISBN 978-0-7923-4543-5.

Castro, R.; Oliver, G.; Withers H. & Almond, P. (1973). *Am. J. Roent.*, Vol.117, p. 182.

Chin, H.; Maruyama, Y.; Patchell, R. & Young, A. (1991). *Nucl. Sci. Appl.*, Vol.4, pp. 261-271.

Debus J. et. al., (1998). Is There a Role for Heavy Ion Beam Therapy, In: (pp. 170-182) *Fast Neutrons and High-LET Particles in Cancer Therapy*, Engenhart R. & Wambersie, A. (Eds.), pp. 170-182, Springer, ISBN 3-540-57632-0.

Fields, P., et al. (1956). Transplutonium Elements in Thermonuclear Test Debris. *Phys Rev*, Vol.102, pp. 180-182.

Fortune, E.; Gauld, I. & Wang, C. (2011). Gamma Dose near a New ^{252}Cf Brachytherapy Source, *Nuclear Technology*, Vol. 175, No. 1, pp. 73-76.

Grundl, J & Eisenhauer, C. (1975). In: *Neutron Cross Sections and Technology* (Proc. Int. Conf., Washington, D.C.), National Bureau of Standards Special Publication 425, U.S. Government Printing Office, Washington, D.C., p. 250.

ICRU Report 26, (1977). Neutron Dosimetry for Biology and Medicine, International Commission on Radiation Units and Measurements, Bethesda, Maryland

ICRU Report 45, (1989). Clinical Neutron Dosimetry, Part I: Determination of Absorbed Dose in a Patient Treated by External Beams of Fast Neutrons," International Commission on Radiation Units and Measurements, Bethesda, Maryland.

Joiner, M. (2002). Particle Beams in Radiotherapy, Chapter 19 in: *Basic Clinical Radiobiology*, Steel, G. (Ed.), 3rd edition, Oxford University Press, Inc., New York, 2002.

Kal, H. (1986). Review of RBE and OER Values for Cf-Neutrons, *Nucl. Sci. Appl.*, Vol.2, pp. 303-316.

Knauer,J.; Alexander C. & Bigelow, J. (1991). Cf-252: Properties, Production, Source Fabrication and Procurement, *Nucl. Sci. Appl.*, Vol.4, pp. 3-17.

Lei, X. et. Al. (2011). Californium-252 Brachytherapy Combined with External-Beam Radiotherapy for Cervical Cancer: Long-Term Treatment Results, *Int. J. Radiation Oncology Biol. Phys.*, doi:10.1016/ijrobp.2010.08.039.

Leibel S. & Sheline, G. (1991). Tolerance of the Brain and Spinal Cord to Conventional Irradiation, Chapter 13 of *Radiation Injury to the Nervous System*, Gutin, P.; Leibel, S. & Sheline, G. (Eds.), Raven Press, New York, ISBN 0881677604.

Lindsley, K. et. al., (1998). Fast Neutrons in Prostatic Adenocarcinomas: Worldwide Clinical Experience, In: *Fast Neutrons and High-LET Particles in Cancer Therapy*, Engenhart, R. & Wambersie, A. (Eds.), pp. 125-136, Springer, ISBN 3-540-57632-0.

Marjina, et al., (1997). In: *Californium-252 Isotope for 21st Century Radiotherapy*, Wierzbicki, J. (Ed.), pp. 115-130, Kluwer Academic Publishers, ISBN 978-0-7923-4543-5.

Martin, R. et. al. (1997). Development of High-activity ^{252}Cf Sources for Neutron Brachytherapy. *Appl. Radiat. Isot.*, Vol.48, No.10-12, pp. 1567-1570.

Maruyama, Y.; Feola J. & Beach J. (1983). *Int. J. Rad. Oncol. Biol. Physics*, Vol.9, p. 1715.

Maruyama, Y. (1986). Californium: New Radioisotope for Human Cancer Therapy, *Endocurietherapy/Hyperthermia Oncology*, Vol.2, pp. 171-187.

Maruyama, Y. et al. (1991). Cf-252 Neutron Brachytherapy Treatment for Cure of Cervical Cancer. *Nucl. Sci. Appl.*, Vol.4, pp. 181-192.

Maruyama, Y & Patel, P. (1991). Dose-effect for Oral Tongue Cancer Using Cf-252 Neutron Brachytherapy. *Nucl. Sci. Appl.*, Vol.4, pp. 251-258.

Maruyama, Y. et al. (1993). Schedule in Cf-252 Neutron Brachytherapy: Complications After Delayed Implant for Cervical Cancer in Phase II Trial. *Am. J. Clin. Oncol.*, Vol.16, No.2, pp. 168-174.

Maruyama, Y. et al. (1997). Californium-252 Neutron Brachytherapy. Chapter 35 of: *Principles and Practice of Brachytherapy*, Nag, S. (Ed.), Futura Publishing Co., Inc., ISBN 0879936541.

McGarry E. & Boswell, W. (1988). NBS Measurement Services: Neutron Source Strength Calibrations, Library of Congress Catalog Card Number: 88-600510, U.S. Government Printing Office, Washington, D.C.

Medvsdev, V. S. et al. (1991). Early and Long Term Results of Interstitial Neutron Cf-252 Brachytherapy of Oral Cavity Mucosal Cancers. *Nucl. Sci. Appl.*, Vol.4, pp. 235-238.

Mosley, W. et al. (1972). Palladium-^{252}Cf Oxide Cermet, and Improved Form for ^{252}Cf Sources. Report DP-MS-72-4, Savannah River Laboratory, Aiken, SC.

Mount, M. (1991). The New brachytherapy and Its Possible Application Using Cf-252, *Nucl. Sci. Appl.*, Vol.4, pp. 397-408.

NDD, Nuclear Decay Data (2011). available from the Brookhaven National Laboratory website: http://www.nndc.bnl.gov/mird.

NIDC, National Isotope Development Center (2011). Newsletter-#01, June, 2011, available from: http://www.isotopes.gov/news/newsletter_archive/newsletter_2011a.pdf.

ORNL Invention Disclosure No. 1289, (2002). under CRADA contract with Isotron Inc.

Paine, C.; Wiernik, G.; Berry, R.; Young C. & Stedeford J. (1976). Physical Dosimetry and Biomedical Aspects of Californium-252, IAEA, Vienna, p. 19.

Patchell, R.; Yaes, R.; Beach, J.; Kryscio, R.; Tibbs P. & Young, B. (1997). Phase-I Trial of Neutron Brachytherapy for the Treatment of Malignant Gliomas, *Brit. J. Radiol.*, Vol.70, pp. 1162-68.

Rivard, M. (1999) Dosimetry for ^{252}Cf Neutron Emitting Brachytherapy Sources: Protocol, Measurements, and Calculations, *Med. Phys.*, Vol. 26, No. 8, pp. 1503-1514.

Rivard, M. (2000) Burst calculations for ^{252}Cf brachytherapy sources, *Med. Phys.*, Vol. 27, No. 12, pp. 2816-2820.

Rivard, M. & Zamenhof, R. (2004) Moderated ^{252}Cf Neutron Energy Spectra in Brain Tissue and Calculated Boron Neutron Capture Dose, *Applied Radiation and Isotopes*, Vol. 61, pp. 753-757.

Rivard, M.; Melhus, C.; Zinkin, H.; Stapleford, L.; Evans, K., Wazer, D. & Odlozilikova, A. (2005). A Radiobiological Model for the Relative Biological Effectiveness of High-Dose-Rate ^{252}Cf Brachytherapy. *Radiat. Res.*, Vol. 164, pp. 319-323.

Sidorchenkov, et al. (1997) Intracavitary Neutron Therapy for Malignant Rectal Tumors Using High Activity Cf-252 Sources, In: *Californium-252 Isotope for 21st Century Radiotherapy*, Wierzbicki, J. (Ed.), pp. 179-198, Kluwer Academic Publishers, ISBN 978-0-7923-4543-5.

Stoddard D. & Hootman, H. (1971). Cf-252 Shielding Guide, Savannah River laboratory Report DP-1246.

Shpiklov V.; Atkochyus, V. & Valuckas, K. (1991). The Application of Cf-252 in Contact Neutron Therapy of Malignant Tumors at the Scientific Research Institute of Oncology in Vilnius, Lithuania, *Nucl. Sci. Appl.*, Vol.4, pp. 419-424.

Stoddard, D. (1986). Historical Review of Californium-252 Discovery and Development. *Nucl. Sci. Appl.*, Vol.2, pp. 189-199.

Stoll, B; Maruyama, Y.; Patel, P. & Kryscio, R. (1991). Cf-252 Neutron Brachytherapy for Advanced Tonsillar-Oropharyngeal Carcinoma. *Nucl. Sci. Appl.*, Vol.4, pp.243-250.

Tacev, T. et al., (1997). In: *Californium-252 Isotope for 21st Century Radiotherapy*, Wierzbicki, J. (Ed.), pp. 83-97, Kluwer Academic Publishers, ISBN 978-0-7923-4543-5.

Tacev, T.; Ptackova, B. & Strnad, V. (2003). Californium-252 Versus Conventional Gamma Radiation in the Brachytherapy of Advanced Cervical Carcinoma - Long-Term Treatment Results of a Randomized Study. *Strahlenther Onkol.*, Vol.179, No.6, pp. 377-384.

Tsuya, A.;Kaneta, K.; Sugiyama, T.; Onai, Y.; Irifune, T.; Uchida, M.; Kaneta, S. & Tsuchida, Y. (1979). *Nippon Acta Radiol.*, Vol.39, p. 32.

Tsuya, A. & Kaneta, K. (1986). Treatment of Cancers of the Tongue and Oral Cavity and Lymph Node metastases with Cf-252 at Cancer Institute Hospital, Tokyo, Japan. *Nucl. Aci. Appl.*, Vol.2, pp. 539-553.

Vallejo, A.; Hilaris B. & Anderson L., (1977) *Int. J. Radiat. Oncol.*, Vol.2, p. 731.

Vtyurin, B. (1982). *Rev of Information, Med., and Pub. Health Svc, Oncology*, Moscow, USSR.

Vtyurin B. & Tsyb, A. (1986). Brachytherapy with Cf-252 in USSR: Head and Neck, GYN and Other Tumors. *Nucl. Sci. Appl.*, Vol.2, pp. 521-538.

Vtyurin, B.; Medvedev, V.; Maksimov, S. & Anikin, V. (1991a). Results of Brachytherapy of Lower Lip Cancer with Cf-252. *Nucl. Sci. Appl.*, Vol.4, pp. 231-234.

Vtyurin, B.; Medvedev, V. & Melin, V. (1991b). Neutron Brachytherapy of Recurrent and Persistent Oral Cavity Tumors with Cf-252. *Nucl. Sci. Appl.*, Vol.4, pp. 239-242.

Vtyurin, B. et. al. (1997). Brachytherapy with Cf-252 for Head and neck Tumors, In: *Californium-252 Isotope for 21st Century Radiotherapy*, Wierzbicki, J. (Ed.), pp. 145-158, Kluwer Academic Publishers, ISBN 978-0-7923-4543-5.

Walsh, R. (1989). Spin-dependent Calculation of Fission Neutron Spectra and Fission Spectrum Integrals for Six Fissioning Systems, *Nucl. Sci. Eng.*, Vol.102, pp. 119-133.

Wang, C. & Zhang, X. (2006). A Nanodosimetry-Based Linear-Quadratic Model of Cell Survival for Mixed-LET Radiations, *Phys. Med. and Biol.*, Vol.51, pp. 6087-6098.

Wang, C.; Zhang, X.; Gifford, I.; Burgett, E.; Adams, V. & Al-Sheikhly, M. (2007) Experimental Validation of the New Nanodosimetry-Based Cell Survival Model for Mixed Neutron and Gamma-Ray Irradiation. *Phys. Med. and Biol.*, Vol.52, pp. N367-N374.

Wang, C. & Kelm R. (2009). Determination of Neutron and Gamma Dose Rates in Water Surrounding a New Interstitial Cf-252 Brachytherapy Source," *Trans American Nuclear Society*, Vol.100, pp. 32-33.

Wanwilairat, S.; Schmidt, R.; Vilaithong, T.; Lorvidhaya, V. & Hoffmann, W. (2000). Measurement of the Dose Components of Fast and Thermal Neutrons and Photons from a 0.1 mg ^{252}Cf Source in Water for Brachytherapy Treatment Planning. *Med. Phys.*, Vol.27, No.10, pp. 2357-2362.

Wierzbicki, J.; Rivard, M. & Roberts, W. (1997). Physics and Dosimetry of Clinical ^{252}Cf Sources, In: *Californium-252 Isotope for 21st Century Radiotherapy*, Wierzbicki, J. (Ed.), pp. 115-130, Kluwer Academic Publishers, ISBN 978-0-7923-4543-5.

Yamashita, H.; Dokiya T.; Yamashita, S.; Ito, H. & Hashimoto, S. (1991). Comments on the Results of Remotely Controlled Afterloading High Dose Rate Therapy of Cancer of the Uterine Cervix Using Cf-252. *Nucl. Sci. Appl.*, Vol.4, pp. 197-199.

Section 3

Practice of Brachytherapy

4

Theoretical, Manufacturing and Clinical Application Aspects of a Prostate Brachytherapy I-125 Source in Brazil

Carlos A. Zeituni et al.*
Instituto de Pesquisas Energéticas e Nucleares,
IPEN-CNEN/SP
Brazil

1. Introduction

One of the treatments commonly applied for the treatment of prostate cancer is the brachytherapy implant with iodine-125 seeds. In this chapter, we will attempt to show how to produce seeds of iodine-125, the main requirements to have a good seed, and also we will try to discuss some issues in using this technique at the clinics.

2. Iodine-125 production

Cancer is one of the worst illnesses in the world and one of the major causes of death in Brazil [Rostelato et. al, 2008]. For this reason, the Brazilian Nuclear Energy Commission (CNEN) started a project to produce some medical radioisotopes to treat cancer. One of the main products is the iodine-125 seeds [Moura et. al, 2010]. This iodine seed can be used to treat several types of cancer: prostate, lung, occular and brain. In the first phase of this project, the iodine-125 will be acquired in international market. As Brazil will construct a new reactor to produce radioisotopes, it is necessary to define how the iodine-125 production will carry out [Cieszykowska et al., 2005; Mathew et. Al., 2002].

The main reaction of this production is the irradiation of the enriched xenon-124 in gaseous form. Natural xenon has only 0.095% of xenon-124. Xe-124 changed to iodine-125 by neutron capture following in two decays:

$$\text{Xe-124 } (n, \gamma) \rightarrow \text{Xe-125m (57s)} \rightarrow \text{I-125}$$

$$\text{Xe-124 } (n, \gamma) \rightarrow \text{Xe-125 (19.9 h)} \rightarrow \text{I-125}$$

The cross section for thermal neutrons (0.0253 eV) makes $\sigma = (165 \pm 20)$ b. The core product, Xe-125m, decays by isomeric transition (100%) with half-life of 57 seconds to Xe-125g which,

* Carla D. Souza[1], Eduardo S. Moura[1], Roberto Kenji Sakuraba[2], Maria Elisa C.M. Rostelato[1], Anselmo Feher[1], João A. Moura[1], Samir Somessari[1] and Osvaldo L. Costa[1]
[1]*Instituto de Pesquisas Energéticas e Nucleares, IPEN-CNEN/SP, Brazil*
[2]*Hospital Israelita Albert Einstein, Brazil*

in turn, decays by electron capture (99.3%) and positron emission (0.7%), with half-life of 16.9 hours, generating iodine-125. Finally, iodine-125 decays by electron capture (100%) for the Te-125. The decay of iodine-125 is accompanied by the emission of photons of 27 keV, 31 keV and 35 keV, average energy of 29 keV. Due to the low average energy of emission, such photons have low penetrating strengh. Iodine-125 has a half-life of 59.408 dias [Zeituni, 2008].

During the production of iodine-125 in a reactor, Iodine-125 produced abosorbs a neutron and it it transformed in iodine-126. Iodine-126 has a half life of 13.1 days and it has severals photons of high energy (388.6 keV, 491.2 keV, 666.3 keV, 753.8 keV, 879.9 keV and 1420.2 keV). Iodine-126 is considered a contaminant [HAN et. Al., 2007]. The research group of IPEN/CNEN-SP decided to usetwo techniques: a batch system and continuousa cryogenic system.

The batch system consists of a sealed capsule (placed in the reactor core for around 64 hours. In this type of production, some iodine-126 is produced and a certain quantity of Xe-124 is not activated. Then it is really important for economic reasons, it uses a system to retrieve the Xe-124, and it needs to wait some days to let the iodine-126 decays. Of course, during this time the iodine-125 decays, too. . Usually, it is necessary to wait around 5 to 7 half-lives to lower the the I-126 contamination level.-After this time, the quantity of iodine-125 is only 50% to 34% imediately after the reactor shutdown.

We calculated the yeald in IPEN/CNENSP reactor, IEA-R1 built in 1956. This reactor outputs around 4.5MW and the thermal flux is 8.10^{12} n/cm^2.s at the xenon/iodine capsule position. After the cycle of irradiation (64 hours) we will have 65 GBq of iodine-125. In this time we will have around 0.26% of iodine-126 produced. After 20 days, the activity of iodine-125 will be 50 GBq and 0.14% of iodine-126. After 40 days, we yield 40 GBq of I-125 with 0.06% of iodine-126.

The second technique used to produce iodine-125 is the continuous production using a cryogenic system with circulating liquid nitrogen. This technique consists in two capsules: one inside the reactor core and the second one out of the neutron flux. These two capsules will be linked with two cryogenic pumps to guarantee that all iodine-125 produced in the core will be take off the reactor core (Figure1). The cryogenic temperature, around 77 K, will turn the iodine liquid but the xenon will be pumped again to the core in gaseous form. After the irradiated cycle, we take up and seal the capsule outside of the core, and consequently yield almost onlyiodine (with a traces of xenon),) and insert another Xenon-124 enriched capsule in the reactor to continue the production. The greatest disadvantage of this technique is the using of two positions in the core of the reactor though Brazil has only one radioisotope reactor producing, andthere is a huge quantity of materials to be produced. Though the current seeds production power in Brazil is only for 3000 seeds per month, though the demand is around 3.5 Ci (around $1.3.10^{11}$ Bq) per month. This batch production may produce a small quantity, but this is more than that of batch production.

Iodine-125 decays by electron capture (EC) and internal conversion to Tellurium-125 as shown in figure 2. Photons with 27 keV, 31 keV and 35 keV (average 29 keV) are issued.

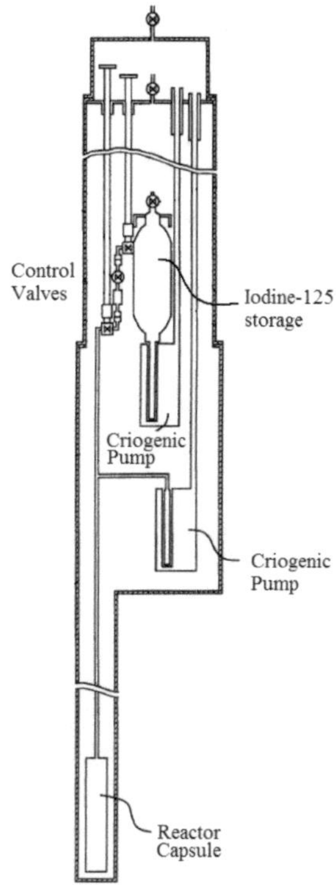

Fig. 1. Iodine-125 production system in cryogenic method [Zeituni et. Al., 2011]. (To publisher, this is too small. To authors, please indicate how to use this unit showing or describing the corresponding reactor part or position: for exapmle, "The reactor capsule part is placed into the reactor core." The figure is not enough to explain how to retrieve the capsule, where to insert next capsule, where to take up the I-125.)

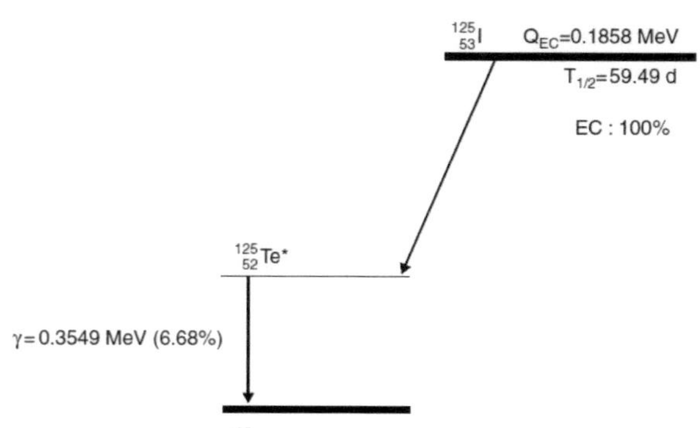

Fig. 2. Iodine-125 decay scheme [Baltas et. Al, 2006]

3. Iodine-125 seeds

Iodine-125 sources are used in the radiotherapy of brain, neck, lung, pancreas and prostate cancers, as well as intraocular tumors (choroidal melanomas and retinoblastomas). They can be used in two different ways - permanent and temporary implants.. Iodine-125 seeds are indicated for the treatment of tumors that have some of the following characteristics: localized, slow growth rate and low to moderate sensitivity. They are also indicated for the treatment of recurrent and residual tumors following a course of external radiation therapy. The mean apparent activity of Iodine-125 seed required to be used in ophthalmic applicators is 20 mCi (740 MBq) and in other applications most often 4-5 mCi (148-185 MBq).

The treatment of prostate cancer with permanent iodine-125 seed implantation has grown dramatically in the world in recent years.

The technique is being used by hospitals and private clinics also in Brazil. Nowadays, the seeds are imported at a minimum cost of US$ 65.00 per one seed, however the high cost makes them impracticable for use in public hospitals. In a typical prostate brachytherapy 80-120 seeds are required. Generally, the seeds are composed of a titanium capsule of 0.8 mm outer diameter, 0.05 mm wall thickness and 4.5 mm in length. The internal structure varies significantly by commercial models. Some constitute of resin or ceramic embedding mixture, and others are deposited in a substrate radio-opaque substrate. Commercially available applicator devices for implantation are disigned for the general dimension.

All seeds are encapsulated in titanium because it is an inert material that does not cause rejection when in direct contact with human tissue and it is classified as biocompatible material. The seed manufacturers in the world are concentrated in the United Kingdom, Belgium and the United States of America and the seeds produced differ in the process used in manufacturing, being unique and protected by patents.

The iodine-125 seeds are classified as sealed radioactive sources as standard *International Standard Organization. Radiation Protection – Sealed Radioactive Sources – General Requirements and Classification ISO-2919.*

The seed is formed by a core that has the radioactive material attached; it is wrapped in a shell of biocompatible material sealed on both sides. A radiological marker must also be inserted. Some examples are materials used in the items below [Rostelato, 2006]:

- UroMed Corporation - Bebig GmbH (Germany) → Symmetra-125.S06: Titanium capsule sealed laser, containing a radio-opaque gold wire inside, and a ceramic layer with iodine-125.
- Best Medical International (USA) → Model 2301: The outer coat consists of double encapsulated titanium, without specifying the type of welding. The interior accommodates a marker of tungsten and iodine-125 adsorbed on an unspecified substrate.
- BARD - SourceTech Medical (USA) → BrachySource STMIodine-125: The capsule is titanium welded by laser equipment. Inside of the seed has a gold wire as a marker, a layer of aluminum and a "coating" of copper. Iodine-125 is deposited in a cylinder of aluminum with a gold core and a layer of nickel.
- Oncura GE Healthcare (USA)→ OncoSeed 6711: Radio-opaque silver core, where iodine-125 is adsorbed and the outer shell is titanium with laser sealing.
- Oncura GE Healthcare (USA)→ OncoSeed 6702: Iodine-125 is adsorbed on ion exchange resin beads. The outer shell is titanium with laser sealing.
- Mentor Corp. - North American Scientific (USA) → IoGold MED3631-A/M: Titanium capsule sealed by laser. Iodine-125 adsorbed onto four resin beads. The capsule shell

contains two inactive gold beads which serve as markers to identify and locate the source.

- Syncor (China) → PharmaSeed BT-125-I: Iodine-125 is adsorbed on palladium wire. The seed is sealed by laser.
- Med-Tec - Implant Sciences (USA) → I-Plant 3500: Iodine-125 is deposited on a ceramic coating. A silver bullet is placed inside the cylinder.
- International Brachytherapy (Belgium) → Intersource Iodine-125L. Iodine-125 is adsorbed on the inorganic matrix ring positioned in the center and beads at the ends of the source. It is used a marker of iridium and platinum.
- UroCor - Mills Biopharmaceuticals (USA) → ProstaSeed I125-SL: Iodine-125 is adsorbed by ion exchange resin in five balls.
- Imagyn Medical - International Isotope (USA) → IsoSTAR 1250I: Iodine-125 is adsorbed on the resin pellets with a diffusion barrier.
- DraxImage - Cytogen (Canada) → BrachySeed LS-1: A glass substrate doped with silver is used. A marker containing 10% platinum is used.

It is important to notice that some seeds shown under here are not produced nowadays and some of them were not commercialized.

4. Dosimetry of Iodine-125 seeds

The dosimetric characteristics of the iodine-125 seeds are performed according to American Association of Physics in Medicine (AAPM) formalism Task Group No. 43(TG-43). This protocol was developed by a committee of brachytherapy dosimetry researchers and implements specific modifications on the physical quantities assigned to the brachytherapy sources, these measurable quantities are used to obtain the dose distribution of the brachytherapy sources [Nath et. al., 1995]. TG-43 underwent an update rendering to the TG-43U1 [Rivard et. al. 2004] and a supplement were published afterwards [Rivard et. al. 2007]. TG-43U1 dose rate calculation will be briefly described here but the details are found in the references [Nath et. al., 1995; Rivard et. al. 2004].

Basically, TG-43U1 recommends that the dosimetry distribution of brachytherapy sources will be performed with experimental and computational methods. Experimental methods usually use thermoluminescent dosimetry methods [Chiu-Tsao et. al. 1990; Nath and Yue 2002; Wallace 2000]. The majority of computational methods uses Monte Carlo (MC) radiation transport codes, for example, EGSnrc, MCNP, Geant, PENELOPE and others [Burns and Raeside 1987; Mobit and Badragan 2003; Thomson and Rogers 2009].

For both experimental and computational methods, the formalism to calculate the dose rates is based in one (1D) and a two-dimensional (2D) configuration. One-dimensional is an approximation for the isotropic point-source. Two-dimensional approximation considers a homogeneous radioactive material distribution along the longitudinal axis of seed active length (region that radioactive material was distributed).

TG-43U1 two-dimensional approximation is described by polar coordinates around a transversal bisector plane of the source. The dose rate surround the source are determined with any chosen point $P(r,\theta)$ on the transversal bisector plane. There is a reference point $P(r_o,\theta_o)$ that provides the reference dose rate $D(r,\theta)$; the reference point has distance (r_0) and angle (θ_0) equal to one centimeter and $\pi/2$ radians, respectively. This reference point was chosen according to traditional dosimetry practices [Nath et. al., 1995]. Fig. 3 illustrates the geometry used for brachytherapy source geometry.

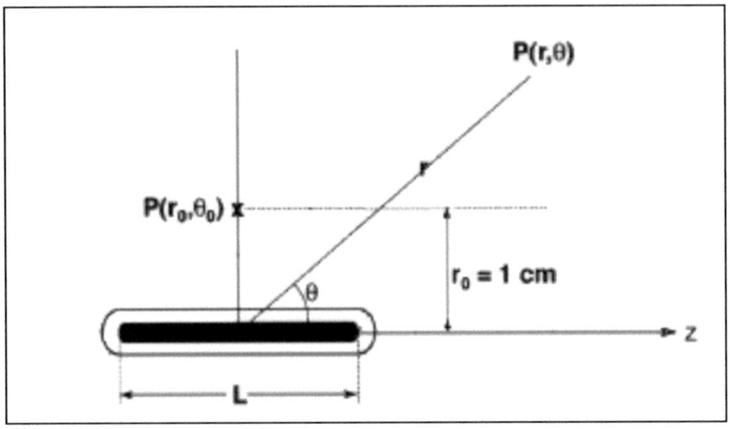

Fig. 3. Geometry used to perform the brachytherapy sources dose measurements, the L denotes the source active length (adapted from Rivard et. al., 2004)

According to the TG-43U1, the general two-dimensional dose rate formalism for brachytherapy sources can be expressed as:

$$\dot{D}(r,\theta) = S_k \cdot \Lambda \cdot \frac{G_L(r,\theta)}{G_L(r_0,\theta_0)} \cdot g_L(r) \cdot F(r,\theta)$$

where S_k is the air kerma strength ($cGy \cdot cm^2 \cdot h^{-1}$ or U), Λ is the dose constant rate (cm^{-2}), $G_L(r,\theta)$ geometry function; $g_L(r)$ Radial dose function; $F(r,\theta)$ anisotropy function. The subscript L denotes the line-source approximation (2D approximation).

Air kerma strength is determined by a single brachytherapy source model, i.e., each brachytherapy seed model has air kerma strength. This quantity is measured by a standard wide-angle free-air chamber (WAFAC) available in National Institute of Standards and Technology (NIST) laboratory [Seltzer et. al. 2003]. Dose constant rate is the ratio of the dose rate in the reference point ($D(r,\theta)$) and the air kerma strength. The geometry function is used to calculate the relative dose due to the spatial distribution of the activity within the source; this function does not consider the scattered and attenuated photons in the source. Scattered and attenuated photons effects are accounts for radial dose function along the transverse seed axis with constant angle ($\pi/2$). Anisotropy function quantify the dose distribution differences due to seed external structure, as the welds on the seeds' extremities [Ravinder et. al., 2004; Nath et. al., 1995]; scattering and absorption effects are included in this function.

Beyond the S_k and $G_L(r,\theta)$ functions, that are obtained from standards laboratories and analytical calculations, respectively, the other functions require the dose rate measurements in the point analyzed. So, the aim of the dosimetric laboratory for brachytherapy sources is to measure the experimental dose in the points-of-interest around the transversal plane seed with the minimum uncertainties. This accuracy is important because the rate dose values in the specific point will be employed in the dosimetric calculations treatment planning. These values obtained from experimental measurements must be compared with the MC simulations, according TG43U1 recommendations.

For experimental studies, the dosimetric laboratory uses LiF:Mg,Ti (TLD-100) thermoluminescent dosimeters purchased by Thermo Scientific (Harshaw-Bricon). TLD-100 is widely used in the brachytherapy dosimetric investigations for many years [Meigooni et. al., 1988; Nath et. al., 1990; Abboud et. al., 2010]. These dosimeters can be presented in many formats and shapes; the micro-cube format with 1 mm x 1 mm x 1 mm was chosen. These dimensions avoid high gradient fields negligible due to the larger dosimeters dimensions.

To obtain precise dose information of the seed we set TLD-100 micro-cubes around of a single iodine-125 seed in a phantom material that represents liquid water. For the iodine-125 seed emitting low energy photons (average energy 28.37 keV) [Usher-Moga et. al., 2008], the Solid Water™ phantom material showed better agreements with liquid water compared with other phantoms materials such as Polymethilmethacrilate (PMMA) and others plastics phantoms [Meigooni et. al., 1988].

Figure 4 shows a Solid Water™ slab of 30 cm x 30 cm x 2 cm in size used in the dosimetric measurements. The slabs were drilled to accommodate the dosimeters in the radial positions in 100 mm of interval, yielding 36 holes per circumference. The hole radial distances along the center of the slab vary from 0.5 cm to 10 cm (Figure 4). The geometry in this seed positioning was to determine the functions of the TG-43 protocol. During the measurement, additional slabs placed above and below the slab containing the seed and dosimeters, these slabs provide backscatter conditions

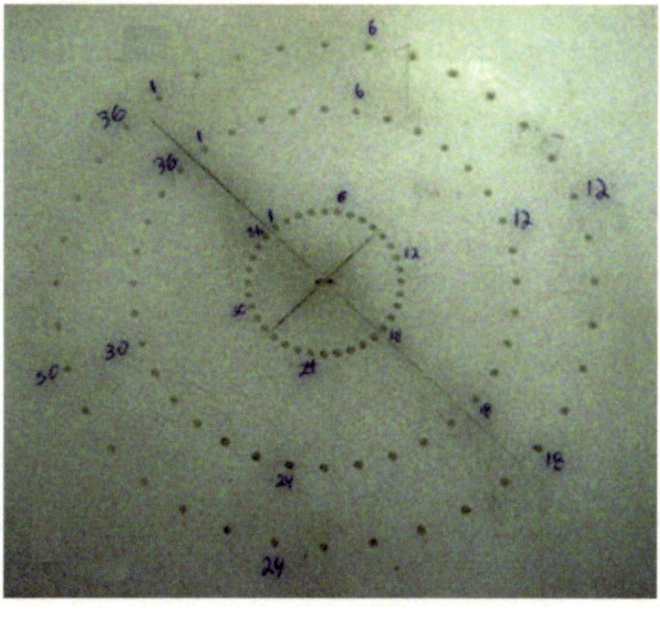

Fig. 4. Solid Water™ slab designed for experimental dosimetric measurements [Zeituni 2008].

Before the dosimetric measurements, the TLD-100 dosimeters were previously selected from a dosimeters' batch. The selected dosimeters represent the minimal uncertainties compared with the overall batch. Difference of the dosimeters masses and edges on the dosimetry structure were some parameters analyzed at the selection [Furreta and Weng, 1998].

All thermoluminescent dosimeters yield different responses of the absorbed dose rate quantity. For dosimeter responses in absorbed dose units, a calibration factor makes useful. This calibration factor is obtained from calibrated beams with a traceable reference dosimetry system, as ionization chambers with special requirements. [Meigooni et. al., 2001; Meigooni et. al., 2006; Abboud et. al., 2010].

The range of the dose delivered in the dosimeters was closed to the absorbed dose levels that will be used in experimental dosimetry routine with brachytherapy sources, it ranges from 10 to 100 cGy. Each dosimeter used in the dosimetry has an individual calibration factor, i.e., one value for the calibration factor to the whole batch was not adopted due to reduce uncertainties propagations on the dose rate calculations. These calibration factors obtained from the x-rays photons were calculated by the averaged measurements originated from the dose levels (10 - 100 cGy).

In summary, the dosimeters that will be measure in a TLD reader will yield thermoluminescent responses in arbitrary units. The intensity of these responses will be proportional to the absorbed dose in the dosimeter. To convert TLD's responses in absorbed dose rates quantities the following formula is applied:

$$\frac{\dot{D}(r,\theta)}{S_k} = \frac{R}{T \cdot S_k \cdot \varepsilon \cdot E(r) \cdot d(T) \cdot F_{lin}}$$

where $\dot{D}(r,\theta)$ is dose rate ($cGy \cdot h^{-1}$) at any point with coordinates (r,θ), S_k : initial iodine-125 seed air kerma strength ($cGy \cdot cm^2 \cdot h^{-1}$) at the start of the measurements; R: the net dosimeter response (nC) of each measured point with backgrounds subtractions,T: irradiation duration (h), ε : calibration factor (nC.cGy-1) for each TLD-100, $E(r)$: a dimensionless correction factor of the TLD-100 between the calibration beam and the iodine-125 photons and the assumed the value was 1.4 for this expression [Abboud et. al., 2010; Reniers et. al., 2002], $d(T)$: a correction factor that explains the source decay during the TLD-100 irradiations and F_{lin} represents a correction for non-linearity in TLD response and was assumed to be unity due to linear response in the doses (10 to 100 cGy) measured with iodine brachytherapy seeds [Meigooni et. al., 2006].

For the TLD-100 measurements a Harshaw 3500 TLD reader is used. The reader operates one dosimeter per load and with continuous flux of pure nitrogen (99,9995% of purity). Nitrogen fluxes are necessary to avoid the signals not delivered from the thermoluminescence dosimeter, as oxygen air induces signals. Heating rate of 10 oC.s-1 and maximum temperature of 260°C are the parameters of TLD-100 measurements used [Thermo Electron Corporation, 2002]. The measurements are performed 20 hours after the exposure; this delay is to avoid unstable glow peak which decays.

TLD-100 dosimeters demand thermal treatments to be reused. Theoretically, the thermal treatment will quench the dose information of previously irradiations. The thermal treatment adopted for the dosimeters is composed by two steps: a) 400ºC for one hour and b) 100ºC for two hours. The thermal variations are performed using the slow cooling rates of approximately 4ºC.min^{-1} [Oster et. al., 2010]. A special oven was designed for the laboratory to attend the slow cooling rates. This oven has coupled cooler that realizes a linear temperature reduction automatically. During the temperature reduction the dosimeters stay inside the oven avoiding drastic changes in the crystal of the dosimeter caused by the gradient temperature.

5. Introduction for hospital methodology

For adenocarcinoma of the prostate conventional treatment options that should be discussed with each patient in this category include radical prostatectomy, external beam radiation therapy, interstitial brachytherapy and watchful waiting, according to the NCI Consensus Conference in 1988 [Lee, 2003].

There are two types of prostate cancer radiation treatments: external and internal (interstitial). Brachytherapy is the interstitial one, meaning treatment is administered "within the tissue." External radiation therapy involves the projection of photon, electron, neutron, or proton beams into the prostate gland from a remote tool called the linear accelerator.

Ionizing radiation is used in the treatment of prostate cancer because exposure to this radiation damages the DNA of cells. Cells will not be damaged unless they attempt to divide. Cancerous cells divide more quickly than healthy cells. Therefore, healthy cells are able to repair damage before undergoing mitosis, while cancerous cells are not. Unfortunately, if the absorbed dose is strong enough, healthy cells will be damaged to the point where they cannot repair themselves before division. Interstitial brachytherapy is able to deliver higher doses of radiation to an area concentrated within the prostate gland [Van Dyk, 1999].

The use of brachytherapy as the sole modality of treatment for early-stage prostate cancer has gained popularity over the past decade due to the advent of the transrectal ultrasound-guided technique (TRUS) and the favorable reports of imaged-based brachytherapy with isotope Iodine-125 (I-125). At the same time, dose escalation 3-dimensional conformal radiation therapy (3D CRT) has revealed promising results, especially for patients with early stage disease [Halperin et. al., 2008].

5.1 Brachytherapy process at the Albert Einstein hospital

Initially the patient is referred to radiotherapy or search for a consultation with a radiation oncologist which defines the treatment plan in conjunction with the urologist involved in the process. The evaluation of radiotherapy will consist of complete history, complete physical examination including digital rectal examination, analysis of all imaging (transrectal ultrasound, computed tomography, magnetic resonance imaging, and / or chest radiography) and laboratory (complete blood count with coagulation, all measures of serum PSA). Additional investigations may be required to meet specific needs of each patient (diabetes, epilepsy, heart disease, renal disease, lung disease, etc). Therapeutic options, including observation, hormone therapy, radical prostatectomy, external beam radiotherapy or implantation of Iodine-125 seeds as well as acute and late side effects, expected or possible, are explained to the patient mainly by the urologist.

5.2 Indication for brachytherapy

The transperineal prostate implants guided by ultrasound is the most indicated and has a great potential to cure the patients diagnosed with prostate cancer with the following conditions:

For treatment alone:

- Stage I or II (the T2aN0M0 T1C);
- Gleason ≤ 6; (no individual values > 3; e.g. 2 +4 +2 or 4 will not be accepted);
- ≤ initial PSA 10 ng/ml;
- Prostate volume ≤ 45 grams;
- Absence of prior transurethral resection (TUR) and
- Absence or small number of microcalcifications.
- Intended dose: 90% of the target volume covered by the CT isodose 160Gy - (144Gy according to recommendations of the TG 43) [Nath et. al., 1995]

Pre Planning and Ultrasound - The examination will be conducted with the equipment Leopard 2001 - B & K, using the ultrasonic transducer type 8558, coupled with the "Stepping Unit" UA1084. The "template" is B & K and the whole system is fixed to the stand Brachystand.

The exam aims to:

- Obtaining images of the entire prostate, with 5mm spacing between slices;
- Transfer these images via cable or VCR to the planning system TherpacPlus (MMS);
- Determination of the volume (in cubic centimeters) of the prostate, using the HWL (Height x Width x Length) x Factor, being the factor of 0.523 for the prostate;
- Identifying in advance the technical feasibility of the implant (anterior urethral defects, pubic arch interference or micro macrocalcifications).

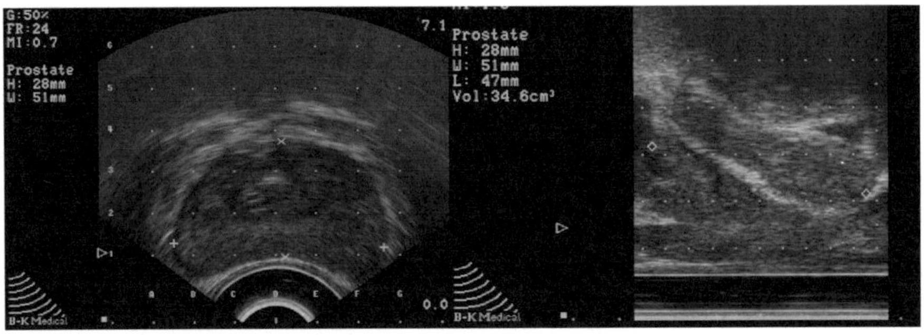

Fig. 5. Prostate volume determination.

Identified clinically relevant changes in the query or ultrasound are discussed with other doctors responsible for feasibility of the procedure (heart disease, diabetes, clotting disorders, use of medications with potential interactions with the procedure used, calcifications, urethral defects, etc.).

5.3 Drawing the boundary of the prostate

In the planning system:

- A new file is created, corresponding to the patient in question;
- A coordinate system is created from the information of the "template" images superimposed on the TRUS (transrectal ultrasound);
- Radiotherapist draws the boundaries of the prostate, seminal vesicles, rectum and ureter in each section of the TRUS;
- Prostate volume is determined by the volumetric reconstruction of the U.S. Distribution and quantity of seeds mCi.

Using system resources Variseed obtain:

- The distribution of seeds, according to the TG 43 protocol and prescribed dose [Rivard et. al., 2004];
- The number of seeds;
- Activity (mCi or units of air kerma) of each seed;
- The number of needles and distribution of seeds per needle.

5.4 Preparation of material for the implant

The team involved in the procedure of radiation therapy (medical, physical, and nursing assistants) handles the preparation of the material to be used in the procedure [Yu et. al., 1999].

The charge physicist needs to control and verify the seeds that reach for each patient. It is recommended that the physicist verify at least 10% of the batch using a well chamber and the measure has a limit of 5% difference between the measured activity and the certificate

The physicist is responsible for taking the seeds in magazines mounted to the center of the second material (Surgical Center Medical Center-CCCM).

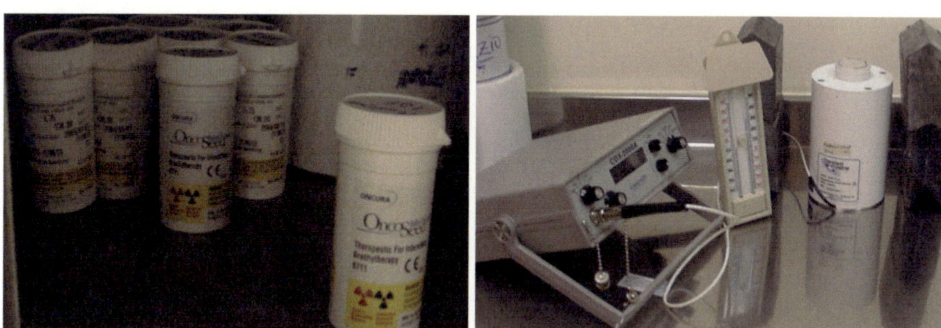

Fig. 6. Dosimetric verification of seeds for implant. Seed container (left). Measurement of each seeds using a well chamber (right).

5.5 Standards for sterilization of seeds

- The seeds are sterilized in an autoclave system the temperature of 121 °C and a pressure of 15 psi for 15 to 30 minutes, or the temperature of 133 °C and pressure of 30 psi for about 3 minutes ("flash").
- After sterilization, using Geiger monitor to check radioactivity inside of the autoclave. Place the test material (seed) in the sterilizer.

- After sterilization, the inside level was checked agian.
- Though the possibility of loss of seeds during sterilization is practically small because of strict arrangement of seeds in the container magazines.[Is this correct?], if it occurs, using the Geiger to direct the search and the use of tweezers to put the seeds in ahead container that bear any responsible for sterilization.
- The physicist takes the seeds to the location of the procedure, where the material will be stored in proper place, behind the screen of lead radiation symbol and more isolated as possible.

5.6 Surgical Room procedures

On the day of the procedure the patient is transferred to the operating room of the radiotherapy center through the nursing assistant of radiation therapy [Prestidge, 2008].

5.7 Positioning the patient on the table

The positioning of the patient on the operating table, and supine and lithotomy position with legs flexed according to the survey pre-planning.

5.8 Anesthesia, monitoring and premedication

The anesthesia may be Epidural, spinal or general. They vary according to the anesthesiologist responsible for assessment and multidisciplinary team member. Aabsolute immobility is essential. The timeis 2-3 hours for one procedure. We set up patient monitorings, and administrate prophylactic intravenous or perineal antibiotic treatment according to official regulation standateds established by Department of Anesthesia, and Office of Infection Control (SCIH), respectively. Saline via the urethral catheter is used to complete the volume, if necessary [Nath et. al., 2009].

5.9 Images of the prostate via ultrasound

- Choose the largest cross section of the prostate, transrectal ultrasound, as the target volume.
- Transrectal ultrasound images has a 5 mm separation of each other, and each image is overlap with the image developed by the planning system. It is used to call this planing image as a template image.
- Adjust the image of the ultrasound so that the "line 1" from the "template" is about 5 mm above the mucosa of the anterior rectal wall, and "column D" centered on the urethra.
- Transfer images to the planning system identifying the prostate, urethra and rectum.
- Revaluate of prostate volume and calculating the number of seeds and needles.

5.10 Insertion of needles and seeds

- Insertion of the needles is by the urologist, according to the shape and size of the prostate and activity of the seeds under the guidance of radiotherapist.
- Identify of each needle in the template and each length to be loaded; Use annotation data file card (We use "Diagram for Seed Implant Prostate with I-125 ultrasound-guided").
- Insert of two needles, via transperineum, approximately 1 to 1.2 cm in the direction of the urethra after 4 and 8 hours as stabilizers.

- Put the needle on the edge first and place them top to bottom.
- Check the positioning of each needle with sagittal images of the ultrasound.
- The needles placed in the periphery are spaced between 0.5 to 1.0 cm and 0.5 cm inside the periphery of the prostate. The typical number of needles in the periphery is 9 to12 needles.
- The needles placed in the "Line 1" (the lowest) are separated by 1.0 cm and about 0.5 cm from the anterior rectal wall mucosa.
- The needles in the central region of the prostate are placed at least 1.0 cm apart from the urethra. The typical number of needles in the central region is 3 to 5.
- The standard distribution of the loads is 75% -80% of the total activity in the periphery and 20% -25% in the center.
- Fill the forms of calculation and distribution of seeds.

5.11 Placement of seeds

- Guided by fluoroscopy and ultrasound at the time the surgery;
- Using the Mick applicator for the loading of individual seeds in each needle according to the pre-planning and the eventual corrections in the time of implantation.
- Check the seed deposition with the help of sagittal ultrasound image to the last needle;
- The physicist and radiotherapist individually confer the number, distribution and spacing of each seed needle immediately prior to their placement, as well as checking the needle by fluoroscopy.
- At the end of seed deposition, potentially cold areas identified by fluoroscopy and ultrasound should be filled with seeds individually.

5.12 Cystoscopy

Realization of cystoscopy is performed by an urologist at the end of the introduction of seeds into the prostate.

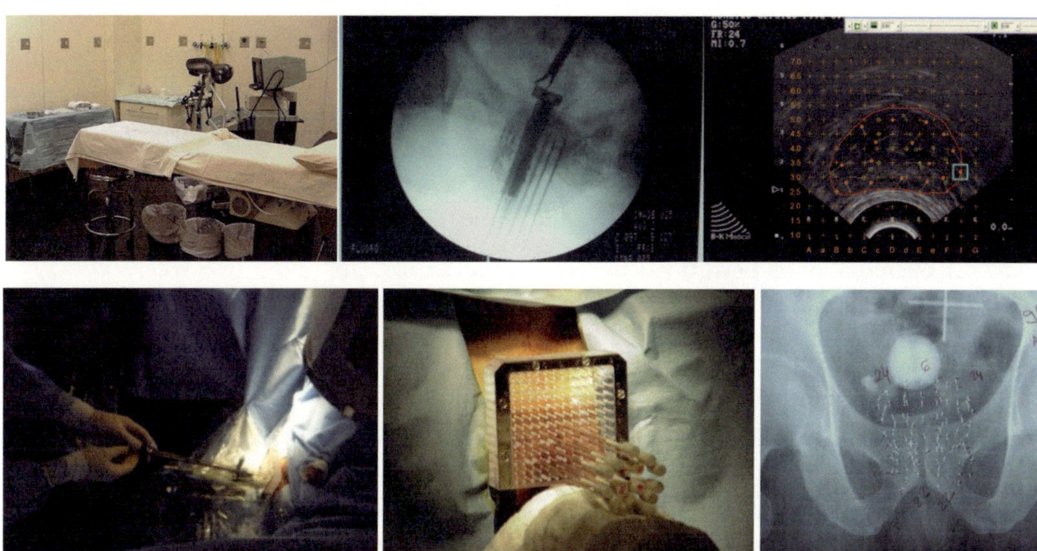

Fig. 7. Insertion of needles and seeds.

5.13 Radiometric survey of the room and the patient

- Counting the number of remaining seeds and deployed to confirm the number of seeds initially loaded in magazines.
- Monitor the environment, professionals and the patient with the monitor Geiger Muller, using the window opened because of the low energy iodine-125.
- The whole procedure for individual and environmental radiological protection is described in "Radiation Protection" section in a published documents by Brazilian Nuclear Energy Commission.

5.14 X-ray control

After the procedure the patient is referred Ximatron CX simulator for the performance of anterior-posterior radiographs and lateral-lateral control and for dosimetric calculations.

5.15 Dose analysis

A quantitative dose analysis must be carried out for each patient post implantation. This statement is based on the premise that it is as important to know and document the dose delivered by a permanent seed implant as by an external beam treatment. The importance of a post implant analysis cannot be overemphasized for the purposes of multi-: institutional comparison, improving techniques, evaluating outcome, and identifying patients who might benefit from supplemental therapy or be at risk for long-term morbidity [Yu et al., 1999].

5.16 Brachytherapy dosimetry

Calculations are performed in accordance with NIST 1999 calibration standards, the point source formalism described in by AAPM Task Group 43, and AAPM Subcommittee Reports [Nath et al., 1995; Rivard et al., 2004; Rivard et al., 2007].

5.17 Prescribed dose

The recommended prescription doses for Iodine-125 are 145 Gy and 110 Gy for monotherapy and boost implants, respectively. The prescription of minimum peripheral dose (mPD) is intended to cover the CTV, and is the reference dose for the treatment.

X-ray CT examination is performed immediately after implant and 3 to 5 weeks after. The patient is scanned in a supine position usually with bladder contrasting. Slices with thickness of 3 mm or less are acquired from 2 cm cephalad to the base of the gland to 2 cm caudad to the apex. All of the seeds used in the implant should be encompassed in the scan. ETVs (-Evaluation Treatment Volume) are determined from this scan, as the location of the urethra and the rectum. Due to the difficulty in CT visualization of the urethra, use of Foley catheterization is strongly recommended. The urethra and the rectum contours are drawn as the outer surface of the Foley catheter and the rectal wall, respectively. The CT images are used to create a post-implant treatment plan (post plan). An AP or anterior oblique pelvic radiograph is used to verify the number of sources and this will be recorded. A surview chest CT image is obtained to check any pulmonary migration of the source.

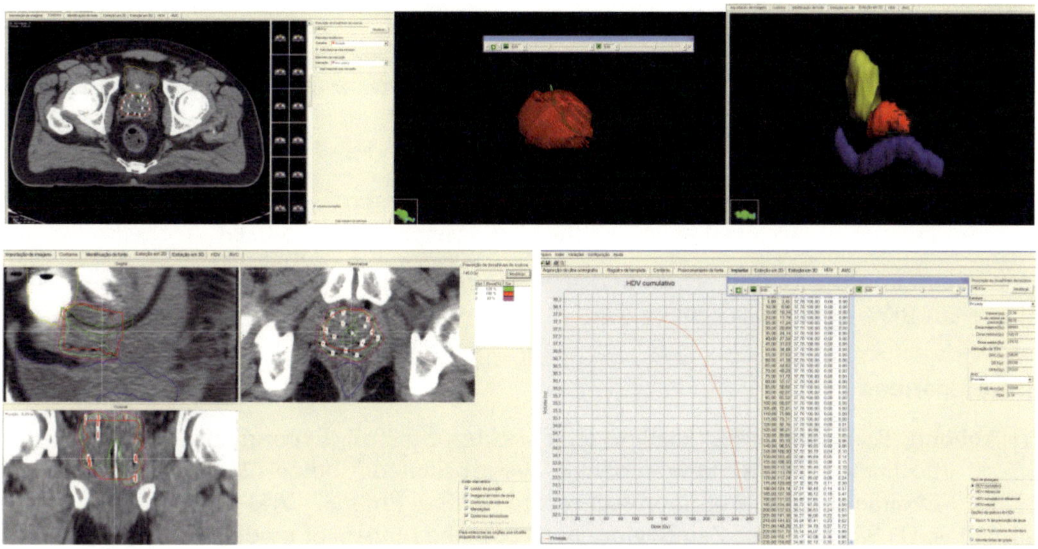

Fig. 8. CT images in the 3D planning system and prostate histogram

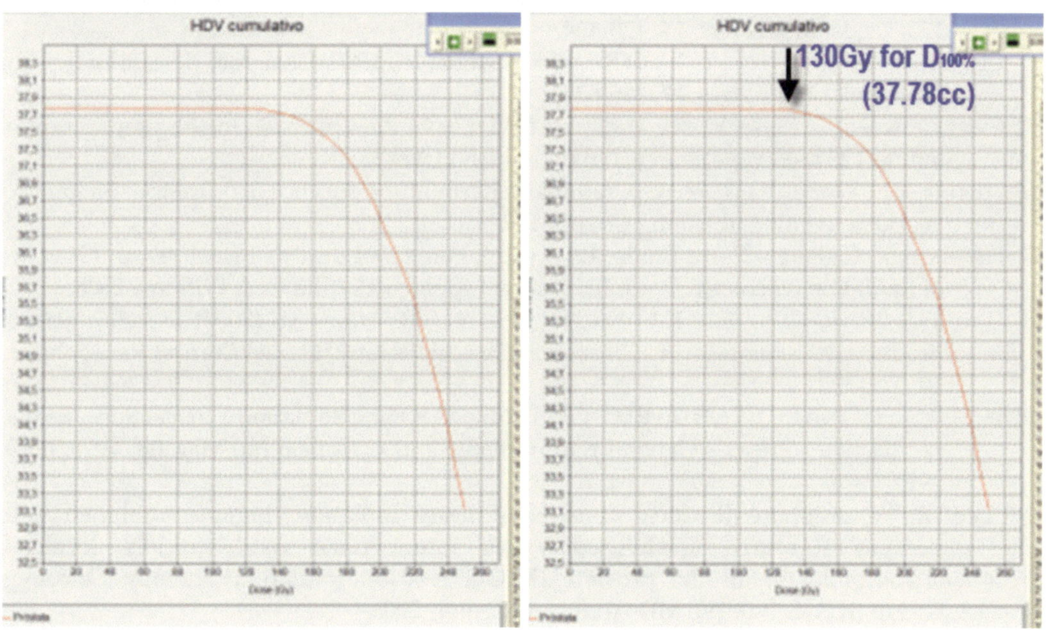

5.18 Dose volume analysis

The planning system facilitates structure-based analysis from axial image sets. This includes evaluation and analysis of isodose curves and generation of Dose-Volume Histograms (DVH). The calculation grid should be set no larger than (2 mm x 2 mm x the axial slice thickness). ABS and ESTRO Guidelines are available [Merrick et al., 2007]. DVH-based analysis must be completed in the post plan evaluation. The following values should be reported:

- Coverage. V100, V90, V80, D90.
- Uniformity. V150.
- Urethra. The maximum dose to the urethra and volume of urethra (in cm^3) that received more than 200% of the prescription dose [U200 (cm^3)].
- Rectum. The outer rectal wall will be contoured behind every axial slice of the prostate.

where Vn is the percentage of the ETV that received at least n% of the prescription dose. Dm is the minimum dose received by m% of the ETV. [Merrick et. al., 2007]:

The maximum dose to the rectum and the volume of the rectum (cm^3) that was received more than 100% of the prescription dose.

6. References

[1] Abboud, F., Hollows, M., Scalliet, P., Vynckier, S., 2010. Experimental and theoretical dosimetry of a new polymer encapsulated iodine-125 source-SmartSeed: Dosimetric impact of fluorescence x rays. Med. Phys. 37, 2054-2062.

[2] Baltas, D.; Sakelliou, L.; Zamboglou, N (2006). The physics of modern brachytherapy for oncology. CRC Press Taylor & Francis Group, ISBN-10: 0750307080, FL/USA

[3] Burns, G. S., Raeside, D. E., 1987. Monte Carlo simulation of the dose distribution around 125I seeds. Med. Phys. 14, 420-424.

[4] Chiu-Tsao, S., Anderson, L. L., O'Brien, K., Sanna, R., 1990. Dose rate determination for 125I seeds. Med. Phys. 17, 815-825.

[5] Cieszykowska, I.; Piasecki, A.; Mielcarski, M. (2005) An approach to the preparation of iodine-125 seed-type source. Nukleonika 50;1:17–22

[6] Furreta, C., Weng, P., 1998. Operational Thermoluminescence Dosimetry. Singapore: World Scientific Publishing Co, pp. 127.

[7] Halperin, E. C.; Perez, C. A.; Brady, L.W. (Ed.). Perez and Brady's Principles and Practice of Radiation Oncology. 5. ed. Baltimore: Lippincott Williams & Wilkins, 2008.

[8] Han, H.S.; Park, Ui-J., Son, K.J.; Lee, J.S.; Hong, S.B.; Ham, S.S (2007) Development of production technology of 125I seed for brachytherapy. J Label Compd Radiopharm; 50: 321–322

[9] INTERNATIONAL STANDARD ORGANIZATION. Radiation protection Sealed radioactive sources - General requirements and classification. Mar. 08, 1995. (ISO-2919)

[10] Lee, R. W. A phase II study of external beam radiation therapy combined with permanent source brachytherapy for intermediate risk clinically localized adenocarcinoma of the prostate (RTOG P-0019). Radiation Therapy Oncology Group (RTOG P-0019), MD Anderson Center, 2003.

[11] Mathew, C.; Majali, M.A.; Balakrishnan, S.A. (2002) A novel approach for the adsorption of iodine-125 on silver wire as matrix for brachytherapy source for the treatment of eye and prostate cancer. Appl Radiat Isot 57:359–367

[12] Meigooni, A. S., Meli, J. A., Nath, R., 1988. A comparison of solid phantoms with water for dosimetry of 125I brachytherapy sources. Med. Phys. 15, 695-701.

[13] Meigooni, A. S., Yoe-Sein, M. M., Al-Otoom, A. Y., Sowards, K. T., 2002. Determination of the dosimetric characteristics of InterSource 125Iodine brachytherapy source. Appl. Radiat. Isot. 56, 589-599.

[14] Meigooni, A. S., Dini, S. A., Awan, S. B., Rafiq, Dou, K., Koona, R. A., 2006. Theoretical and experimental determination of dosimetric characteristics for ADVANTAGETM Pd-103 brachytherapy source. Appl. Radiat. Isot. 64, 881–887. 14 - Kennedy, R. M., Davis, S. D., Micka, J. A., DeWerd, L. A., 2010. Experimental and Monte Carlo determination of the TG-43 dosimetric parameters for the model 9011 THINSeedTM brachytherapy source. Med. Phys. 37, 1681-1688.

[15] Merrick, G. S.; Zelefsky, M. J.; Sylvester, J.; Nag, S.; Bice, W. American brachytherapy society prostate low-dose rate task group. ABS, 2007.

[16] Mobit, P., Badragan, I., 2003. Response of LiF-TLD micro-rods around 125I radioactive seed. Phys. Med. Biol. 48, 3129-3142.

[17] Moura, J.A.; Moura, E.S; Sprenger, F.E.; Nagatomi, H.R.; Zeituni, C.A.; Feher, A.; Manzoli, J.E.; Souza, C.D.; Rostelato, M.E.C.M.; (2010) Tubing decontamination during the leak test of iodine-125 seeds. Nukleonika; 55(3):409-413.

[18] Nath, R., Meigooni, A. S., Meli, J. A., 1990. Dosimetry on transverse axes of 125I and 192Ir interstitial brachytherapy sources. Med. Phys. 17, 1032-1040.

[19] Nath, R., Anderson, L.L., Luxton, G., Weaver, K.A., Williamson, J.F., Meigooni, A.S., 1995b. Dosimetry of interstitial brachytherapy sources: recommendations of the AAPM Radiation Therapy Committee Task Group No. 43. Med. Phys. 22, 209-234.

[20] Nath, R., Yue, N., 2002. Dosimetric characterization of an encapsulated interstitial brachytherapy source of 125I on a tungsten substrate. Brachytherapy 1, 102-109.

[21] Nath, R.; Bice, W. S.; Butler, W. M.; Chen, Z.; Meigooni, A. S. Narayana, V.; Rivard, M. J.; Yu, Y. AAPM recommendations on dose prescription and reporting methods for permanent interstitial brachytherapy for prostate cancer: Reportof Task Group 137. Med. Phys. 36 (11), 2009.

[22] Oster, L., Horowitz, Y., Zlotopolsky, S., 2010. Investigation of the optical absorption charactheristics of slow-cooled LiF:Mg,Ti (TLD-100). Radiat. Meas. 45, 347-349.

[23] Prestidge, B. R.; Amin, M.; Sanda, M.; Bruner, D. W.; Hartford, A.; Bice, W.; Michalski, J.; Ibbot, G. A phase III study comparing combined external beam radiation and transperineal interstitial permanent brachytherapy with brachytherapy alone for selected patients with intermediate risk prostatic carcinoma (RTOG-0232). Radiation Therapy Oncology Group (RTOG 0232), MD Anderson Center, 2008.

[24] Reniers, B., Vynckier, S., Scalliet, P., 2002. Dosimetric study of a new palladium seed. Appl. Radiat. Isot. 57, 805-811.

[25] Rivard, M. J., Coursey, B. M., DeWerd, L. A., Hanson, W. F., Huq, M. S., Ibbott, G. S., Mitch, M. G., Nath, R., Williamson, J. F., 2004. Update of AAPM Task Group No. 43 Report: A revised AAPM protocol for brachytherapy dose calculations. Med. Phys. 31, 633-674

[26] Rivard, M. J., Butler, W. M., DeWerd, L. A., Huq, M. S., Ibbott, G. S., Meigooni, A. S., Melhus, C. S., Mitch, M. G., Nath, R., Williamson, J. F., 2007. Supplement to the 2004 update of the AAPM Task Group No. 43 Report. Med. Phys. 34, 2187-2205.

[27] Rostelato, M. E. C. M., 2006. Estudo e Desenvolvimento de uma nova Metodologia para Confecção de Sementes de Iodo-125 para Aplicação em Braquiterapia. Ph. D. Thesis, University of São Paulo.

[28] Rostelato, M.E.C.M.; Rela, P. R.; Zeituni, C.A.; Feher, A.; Manzoli, J.E.; Moura, J.A.; Moura, E.S.; Silva, C.P.G. (2008) Development and production of radioactive sources used for cancer treatment in Brazil. Nukleonika; 53(Supplement 2):S99-S103

[29] Seltzer, S. M., Lamperti, P. J., Loevinger, R., Mitch, M. G., Weaver, J. T., Coursey, B. M., 2003. New National Air-Kerma-Strength Standards for 125I and 103Pd Brachytherapy Seeds. J. Res. Natl. Inst. Stand. Technol. 108, 337-358.

[30] Thermo Electron Corporation, 2002. Harshaw Standard TTP Recommendations Technical Notice. Publication Number: DOSM-0-N-1202-001.

[31] Thomson, R. M., Rogers, D. W. O., Monte Carlo dosimetry for 125I and 103Pd eye plaque brachytherapy with various see models. Med. Phys. 37, 368-376.

[32] Usher-Moga, J., Beach, S. M., DeWerd, L. A., 2008. Spectroscopic output of 125I and 103Pd low dose rate brachytherapy sources. Med. Phys. 36, 270-278.

[33] Van Dyk, J. (Ed.). The modern technology of radiation oncology - A compendium for medical physicists and radiation oncologists. Madison: Medical Physics Publishing, 1999.

[34] Wallace, R. E., Empirical dosimetric characterization of model I125-SL 125Iodine brachytherapy source in phantom. Med. Phys. 27, 2796-2802.

[35] Yu, Y.; Anderson, L.L. Li, Z.; Mellenberg, D. E.; Nath, R.; Schell, M. C.; Waterman, F. M.; Wu, A.; Blasko, J. C. Permanent prostate seed implant brachytherapy: Report of the American Association of Physicists in Medicine Task Group No. 64. Med. Phys. 26 (10), 1999.

[36] Zeituni, C. A., 2008. Dosimetria de fontes de iodo-125 aplicadas em braquiterapia. Ph. D. Thesis, University of São Paulo.

[37] Zeituni, C.A.; Rostelato, M.E.C.M.; Son, K.J.; Lee, J.S.; Costa, O.L.; Moura, J.A.; Feher, A.;Moura, E.S.; Souza; C.D.; Mattos, F.R.; Peleias Jr., F.S.; Karam Jr., D. (2011) Production of iodine-125 in nuclear reactors: advantages and disadvantages of production in batch or continuous production in cryogenic system. Proceedings of International Conference on Developments and Applications of Nuclear Technologies - NUTECH-2011.

5

High Dose Rate Endobronchial Brachytherapy in Lung Cancer in Patients with Central Airway Obstruction

Alejandro B. Santini[1], Benjamín G. Bianchi[1], Dionis D. Isamitt[2], Claudia C. Carvajal[3] and Gonzalo P. Silva[3]
[1]*National Cancer Institute (INC), Santiago,*
[2]*National Thorax Institute (INT), Santiago,*
[3]*Los Andes University*
[1,2]*Chile*
[3]*Colombia*

1. Introduction

The term brachytherapy derives from the Greek "brachys", which means "short" or "close". It is the branch of radiotherapy in which the radioactive source is located near the therapeutic target. The reported first use of endobronchial brachytherapy (EBBT) done by Yankauer in 1992[1],. He inserted Radon seeds directly into lung tumors though a rigid bronchoscope.

Recent technological development of brachytherapy equipment with small and high dose rate sources, clinical application of EBBT became safe, fast, remote, precise, and less invasive (Fig 1), expanding its indication and feasibility in symptomatic palliation and curative treatment.

One of the major role of EBBT is palliation of symptoms caused by endobronchial cancer ingrowth. Boost EBBT to endobronchial gross tumors combined with external beam radiotherapy (EBRT) provides not only palliative but curative possibilities[2]. In small endobronchial tumors EBBT is used as definitive curative treatment[3].EBBT is also used for non-oncologic pathologies[4,5]. The majority of non-small cell lung cancer (NSCLC)found at loco-regionally advanced stage and frequently associated with bronchial obstruction Various endoscopic techniques available today are including cryotherapy, stent, laser, photodynamic, and EBBT[6,7]. Among these EBBT in the only one that provides biologically tumoricidal effect keeping the normal tissue structure as is.

The National Institute for Cancer in Santiago, Chile (Instituto Nacional del Cáncer de Santiago de Chile) has had EBBT HDR since October 2004, and began treating patients in March 2006, along with the Bronchoscopy department in Chest National Institute. The purpose of this work is to analyze of the result obtained in our center in the palliation of symptoms related to the tumor obstruction of the airway, and literature review.

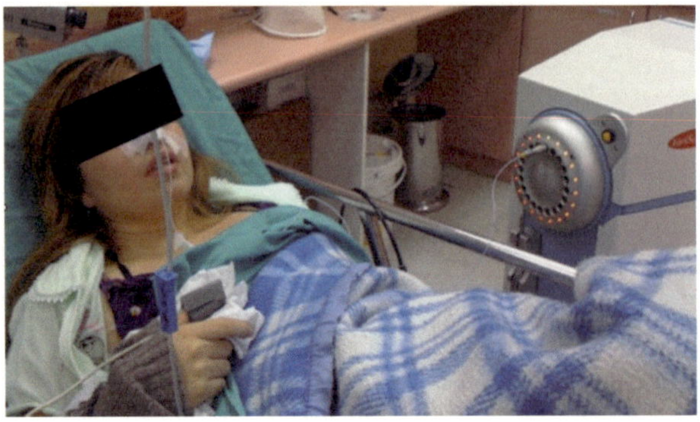

Fig. 1. Varisourse Brachytherapy equipment (patient with a catheter in airway during treatment)

2. Patients, materials and methods

From March 2006 to November 2009, 27 patients were treated with 82 HDR-EBBT procedures as palliative treatment for advanced cancer.

The indication criteria for included: Being able to tolerate decubitus supine position that allows fiberrbronchoscopy examination(FBC), there were clear evidence of tumors with intraluminal component (with or without extrabronchial component), life expectancy of over 3 months, no immediately life-threatening airway obstruction , and healthy coagulating condition. All patients were clinically assessed and the symptoms were evaluated according to Speiser and Spratling[8], recommended by American Brachytherapy Society (ABS), (table 1), prior to and one week after the end of the treatment.

Premedication including included sedatives, and local endobronchial anesthesia which were commonly used in usual FBC.

We used two equipment:

a. Video bronchoscope Fujinon, model EB-270S working channel of 2.0 mm, flexible tube of 4.9 mm in diameter.
b. Fibrobronchoscope Olympus, model BF-TE2 working channel of 2.8mm, flexible tube of 4.9 mm in diameter.

During FBC, the physician evaluated the percentage of luminal obstruction of the air way (table 2), repeating these evaluations after the following FBC.

At the first FBC, the extension of the lesion was measured: the anatomic potion of thee carina was used as reference point to measure lesion extension towards cranium or caudal. After confirmation of the precise target location that was recognized as clearly limited or demarcated, one or two endobronchial brachytherapy applicator catheters (Varian (metal-tipped 4.7, FR-150 cm long (PTFE) R)) were installed ,, with a radiopaque terminal that keeps its position by means of nasal fixing. After insertion of the catheter, or in advance, we used of codeine for managing cough caused by the dwelling catheters in the airway.

The HDR-EBBT planning was simulated on computer associated tomography (CAT) images. using planning computer (Brachyvision system , Varian Medical Systems), which was transferred to are mote high loading dose (iridium-192 source) brachytherapy equipment (VariSource iX, Varian Medical Systems). We prescribed 1 to 4 fractions of 7 to 7.5 Gy (range of total, 7 - 30Gy)at 1 cm depth point from the endobronchial surface-covering the macroscopic lesion with a safety margin of additional 1 cm (Figure 2). The fractional interval between brachytherapy session was 1 week. In patients with significant or complete bronchial obstruction, we applied electrofulguration before installing the catheter, or installed the catheter through the tumor stenosis/obstruction into the peripheral end. Both the catheter installation procedure by BFC and HDR-EBBT were performed in an outpatient setting. Patients were followed up and evaluated for effect and complications at next week of the last brachytherapy session and at the second month. .

Grade	Description
Dyspnea	
0	Without dyspnea
1	Dyspnea in moderate effort
2	Dyspnea in normal activity
3	Dyspnea in rest
4	Requires oxygen
Cough	
0	Without cough
1	Intermittent, no medication required
2	Intermittent, no narcotics required
3	Constant, narcotics required
4	Constant, not improved with narcotics
Hemoptysis	
0	Without hemoptisis
1	Less than twice a week
2	More than twice a week, but less an everyday
3	Daily, red bright blood or clot
4	Hemoglobin decreases and/or hematocrit in more than 10% or more than 150 ml. Requires to be admitted in hospital or transfusion of more than 2 U of red blood cells.

Table 1. Sympton's scale of American Brachytherapy Society (Speiser and Spratling)

Grades	Obstruction
G0	Without evidences of obstruction
G1	Equivalent or less than 25% of obstruction
G2	Obstruction between 25-75%
G3	Obstruction more than 75%
G4	Full obstruction

Table 2. Grade of obstruction

3. Results

Patient consisted of 15 men and 12 women and their age ranged 32-85 years old. The most frequent histologic cancer type was NSCLC, with the main subtype being squamous carcinoma (40%). 7 patients (26%) were previously treated with palliative radiotherapy:(5 patients received 40 Gy in 20 fractions, one received 30 Gy in 10 fractions, one 20 Gy in 5 fractions and the other one 8 Gy in 1 fraction.. At the time of EBBT no patients were receiving concomitant chemotherapy, while a few received chemotherapy previously for relapsed tumor. Table 3 summarizes the characteristics of the 27 patients who received HDR-EBBT.

Lesion location, extent and degree of airway obstruction are described in Table 3 and Figure 3.

With this treatment, a significant improvement was observed in every evaluated symptom (hemoptysis, cough, dyspnea and obstruction). Table 4 compares symptomatic scores before and 1 week after treatment. One patient did not complete treatment for personal reasons, and another one died secondary to a hemorragic complication from the contralateral bronchial tree. Thus 25 patients were evaluable for dyspnea, hemoptysis and cough. 22 patients underwent bronchoscopic reevaluation for degree of obstruction after treatment: One patient with significant clinical improvement refused the reevaluation.

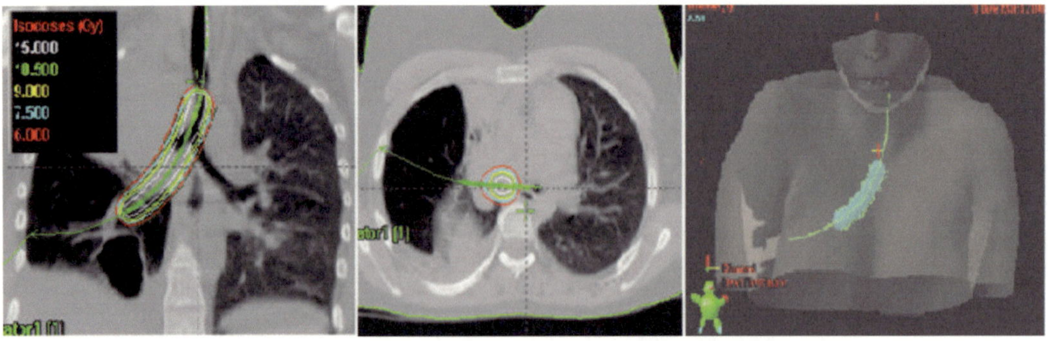

Fig. 2. Dose distribution of a treatment planning to main left bronchus lesion causing left superior lobar atelectasis. Thick green line: catheter in treatment position: Cayn cloured contur between red and yellow indicates 100% of the prescribed dose .

Variables	Number (%)
Age (years)	
Mean ± standard deviation	63,07 ± 12,95
Sex	
Male	15 (56%)
Female	12 (47%)
Diagnosis	
Lung Cancer	21(78%)
Amyloidosis	1 (3%)
Esophageal Cancer	1 (3%)
Renal cancer	3 (12%)
Bronchial carcinoid	1 (3%)
Cancer Histology	
Adenocarcinoma	10 (37%)
Epidermoid	11 (40%)
Undifferentiated NSCLC	1 (3%)
Bronchial carcinoid	1 (3%)
CP large cell	1 (3%)
Small cell CP	1 (3%)
Cancer Stages	
IIIA	1 (4%)
IIIB	7 (28%)
IV	16 (64%)
Relapses	1 (4%)
Location	
Right bronchus	18 (66%)
Left bronchus	11 (40%)
Carina	14 (51%)
Trachea	4 (14%)
Palliative external beam radiotherapy (overprint)	8 (30%)

Total number of staged patinets is 26 because eliminating a patients with benign concurrent disease.
Most of the patients had more than one location.

Table 3. This table summarizes the characteristics of the 27 patients who underwent endobronchial brachytherapy high doses (HDR-EBBT)

Variable	Degree of symptoms	Prior to the EBBT n patients (%)	Post to the EBBT n° patients (%)
Dyspnea	G0	3 (11,1%)	10 (37%)
	G1	2 (7,4%)	13 (48,1%)
	G2	13 (48,1%)	2 (7,4%)
	G3	8 (29,6%)	0
	G4	1 (3,7%)	0
Hemoptysis	G0	4 (14,8%)	25 (92,6%)
	G1	3 (11,1)	0
	G2	9 (33,3%)	0
	G3	11 (40,1%)	0
	G4	0	0
Obstruction	G0	1 (3,7%)	9 (33,3%)
	G1	4 (14,8%)	12 (44,4%)
	G2	11 (40,07%)	1 (3,7%)
	G3	8 (29,6%)	0
	G4	3 (11,1%)	0
Cough	G0	3 (11,1%)	9 (33,3%)
	G1	2 (7,4%)	13 (48,1%)
	G2	9 (33,3%)	3 (11,1%)
	G3	13 (48,1%)	0
	G4	0	0

Table 4. Patient symptoms according to the Speiser-Spratling scale before and after HDR-EBBT

Symptomatic improvement was observed in all hemoptysis condition, 40% of dyspnea, and 25% coughing condition after treatment. After treatment, no patient had G3 or G4 symptoms.

The EBBT procedure was well tolerated, with no acute complications registered. Two months after treatment, two patients (7%) had significant hemoptysis and bronchoscopic examination revealed these bleeding were originated from different locations.. One patient discontinued treatment because of personal choice.

The average treated volume (100% isodose) was 38.1 cm^3(SD ± 15.4)(this shows the smallness of the irradiated tissue volume with this technique)

4. Discussion

The HDR-EBBT provides rapid and significant symptomatic improvement of related to main airway obstruction when the selection of patients is adequate. Our results are consistent with that reported previously (Table 5). These excellent results, with the progress in radioisotope security issue and the outpatient treatment possibility owing to recent decade's technological advances, have made EBBT one of the major applications of brachytherapy in several nations[16].

Currently, EBBT is a part of the therapeutic arsenal in interventional bronchoscopy, being complementary to other techniques such as cryotherapy, laser, photodynamic, stents and argon plasma[17].

EBBT Indications

There are several comprehensive publications that develop the indications and treatments techniques[6, 18.19].

According to the recommendations of the ABS (American Brachytherapy Society)[18] and the ESTRO (European Society of Therapeutic Radiology and Oncology)[19] indications could be divided into: Palliative treatment, healing and non-oncologic pathology.

1. Palliative treatment

Patients with large endobronchial tumors which cause symptoms such as dyspnea due to obstruction, cough, bleeding, or post obstructive pneumonia, represent the most common EBBT indication. Contrary, those tumors that cause obstruction mainly due to extrinsic airway compression would not be candidates for this treatment. Patients with critical obstruction and airway compromise are not recommended for exclusive use of this technique, but it can be used after initial treatment with laser, cryotherapy, stents or electrocautery[20, 21] with great results and ostensible improvement in quality of life.

According to Nag[18], the EBBT might be more effective than a two or three weeks treatment with external beam radiotherapy (EBRT), indicated at first instance, in patients with life expectancy greater than three months that are not candidates for surgery or EBRT due to poor lung function after radiation therapy. Kelly et al, from MD Anderson, discussed their 10 years' experience with EBBT[11]. These authors treated a group of patients with the worst prognosis. Given that, two thirds had already received palliative EBRT. However, they got a 66% improvement of symptoms. Escobar-Sacristán et al, from the Military Hospital of Madrid, reported 85 patients with symptomatic advanced lung cancer who received 288 EBBT applications with 85% partial improvement in symptoms (cough, dyspnea, hemoptysis, and obstruction) and 60% complete responses using endoscopic EBBT[22].

The treatment schedule, i.e., the dose and the number of applications vary in different sites and authors as shown in Table 5. When treatment is performed exclusively, most publications have suggested dose and fractionation schemes between 15 and 35 Gy in one to five applications, calculated within 1 cm from the source. However, more recent studies recommended not exceeding 30 Gy with EBBT[19]. The ABS recommends 3 fractions of 7.5 Gy, two of 10 Gy or four of 6 Gy. These fractions have the same radiobiological equivalent according to the linear quadratic model[23], and the results are very similar[7,13,24-26].

Authors	N° of patients	Dose and fraction	Results
Macha et al[9]	365	5 Gy/3-4f	66% obstructive symptoms improvement
Muto et al[10]	320	GROUP A 10 Gy/1f GROUP B 7 Gy/2f GROUP C 5 Gy/3f	94% hemoptysis and 90% dyspnea reduction, 70% performance status improvement
Kelly et al[11]	175	15 gy/1f	66% symptoms and 85% endoscopic lesions improvement
Petra et al[12]	67	5-7 Gy/1-2f	90% symptoms and 85% endoscopic lesion reduction
Celebioglu et al[13]	95	7-10 Gy/1-3f	Significant reduction of symptoms, hemoptysis, dyspnea and cough
Delclos et al[14]	81	15 Gy/1f	62% symptomatic improvement
Anacak et al[15]	28	7 Gy/3f	43% in cough, 80% in dyspnea and 95% in hemoptysis Improvements
Speicer and Spratling[8]	151	7 Gy/2-3f	99% of hemoptysis, 86% of dyspnea and 85% of cough improvements

Table 5. Published results of HDR endobronchial brachytherapy as palliative treatment

2. Curative treatment

EBBT has been used in curative treatment of lung cancer, both in the exclusive treatment of early tumors with EBRT as overprint (boost or reirradiation)[2, 3]. About the first indication, Marsiglia et al, reported a two year survival rate of 78% in 34 patients with small tumors without evidence of spread that were treated only with HDR-EBBT with a 30 Gy dose in 6 fractions[3].Hennequin et al, in a series of 106 patients with localized tumors that were not candidates for surgery or EBRT, obtained a specific 3 and 5 years survival of 60% and 50% respectively[2].The ABS recommended a healing (palliative) dose is 3 fractions of 5 to 7.5 Gy (total 15 - 22.5Gy) when the patient already received 60 Gy pretreatment with EBRT, and 5 to 6 fractions of 5 to 7 Gy (total 25 - 42 Gy) if this treatment is exclusively used [17].

3. Non-oncologic pathology treatment

Several authors have published interesting experiences regarding the use of EBBT in non-malignant pathology. This treatment modality has been used successfully in patients with granulomatous proliferation after metal stent placement because of non-malignant airway stenosis, or pulmonary transplant patients[4,5].

In patients who have not received EBRT it is feasible to use the combination of EBBT and EBRT in order to improve results. In a randomized study, the combination of those treatments proved to be more effective than single-EBRT in symptomatic improvement[27]. Mantz and colleagues, obtained a significant increase in local control when using a EBBT boost or overlay to EBRT compared to exclusively use of EBRT (58 versus 32% at 5 years, respectively)[28].

We understand that the EBRT is preferred as a palliative treatment when the airway compression is extrinsic, when EBBT is not indicated. In patients with tumors, with an intrinsic component in the airway lumen, Kelly and colleagues from MD Anderson, say that one of the advantages of HDR-EBBT would be the shorter treatment time[11].

5. Conclusion

The HDR-EBBT is a useful technique in the symptomatic treatment of ambulatory patients with central airway obstruction. The procedure is well tolerated and effective with low complications rate. Main indications are patients with advanced lung cancer, where the objective is only palliative. It must also be considered in situations of incipient tumors for patients with contraindications for surgery or EBRT, boosted after a first treatment with EBRT or in benign disease with endobronchial or endotracheal scar proliferation.

6. References

[1] Yankauer S. Two cases of lung tumour treated bronchoscopically. NY Med J 1922;115:741-742

[2] Hennequin C, Bleichiner O, Trédaniel J y col. Long term results of endobronchial brachhytherapy: a Curative treatment? Int J radiat Oncol Biol Phys 2007;67(2);425-30.

[3] Marsiglia H, Baldeyrou P, Larigau E, Brioth E, Haie C, Le chevalier L y col. High Dose Rate Brachytherapy as sole Modality for early stage Endobronchial carcinoma. In J Radiat Oncol Biol Phys 2000:47(3):665-672

[4] Kramer MR, Kats A, Yarmolowsky A y col. Successful use of high dose rate brachytherapy for non-malignant bronchial obstruction. Thorax 2001;56:415-416

[5] Brenner, M. R. Kramer, A. Katz, R. Feinmesser, y col. High Dose Rate Brachytherapy for Nonmalignant Airway Obstruction: New Treatment Option. Chest, 2003; 124(4): 1605 - 1610.

[6] Vergnon JM, Huber RM and Moghissi K. Place of cryotherapy, brachytherapy and photodynamic therapy in therapeutic bronchoscopy of lung cancers. Eur Eespir J 2006;28;200-18

[7] Cavaliere S, Ventura F, Foccoli P y col. Endoscopic treatment of malignant airway obstruction in 2008 patient CHEST 1996:110.1536-42

[8] Speiser BL, Sparting L. Remote afterloading brachytherapy for local control of endobronchyal carcinoma. Int J Radiat Oncol Biol Phys 1993;25:579-87

[9] Macha HN, Wahlers B, Reichle C y col. Endobronchial radiation therapy for obstructing malignancies. Ten years experience with IR192 hiagh-Dose-radiation Brach therapy afteroading technique in 365 patients. Lung 1999;173:274-80

[10] Muto P, Ravo V, Pannelli C y col. High-Dose-rate brachytherapy of bronchial cancer: treatment optimization using thee schemes of therapy. The Oncology 2000;5:209-14

[11] Kelly J, Delclos M, Morice R Y col. High-Dose.rate Endobroncial Brachytherapy afterloading palliative symptoms due to airway tumor. The ten years MD Anderson Cancer Center experience Int J Radiat Oncol Biol Phys 2000;48(3):697-702

[12] Petera J, Spasora I, Neumanova R y col. High-Dose-Rate Intraluminal brachytherapy in treatment of malignant airway obstruction. Neoplasma 2001;48:148-153
[13] Celebioglu B, Ural Gurkan O, Erdogan S y col. High dose rate endobronchyal brachytherapy effectively palliates symptoms due to inoperable lung cancer. Jpn J Clin Oncol 2002.32(11);443-48
[14] Delclos ME, Komaki R, Morice RC y col. Endobronchyial brachytherapy with High-Dose-Rate Remote afterloading for recurrent endobroncyal leson. Radiology 1996:201:279-282
[15] Anacak Y, Mogulkoc N, Ozkok S y col. High dose rate endobronchial brachitherapy in combination with external beamradiotherapy for stage III non-smal cel lung cáncer. Lung cancer 2001.34:253-259
[16] The Royal College of Radiologist. The Role and development of brachitherapy service in United Kingdom. London : the Royal College of Radiologyst, 2007
[17] Janssen JP, Noppen M and Rabe KF. Place of cryotherapy, brachytherapy and photodynamic therapy in therapeutic bronchoscopy of lung cancer. Eur Respir J 2006;28:200-18
[18] Nag A. Brachytherapy for carcinoma of the lung. Oncology 2001;15(3): 371-81
[19] Van Limbergen E and Pötter R. Bonchus cancer in: Gerbaulet A, Potter R, Van Limbergen E, Mazeron JJ and Meertens H: GEC - ESTRO handboock of brachytherapy. ACCO, Belgium 2002
[20] Freitag L, Emst E, Thomas M y col. Sequential Photodinamic therapy and HDR brachytherapy for endobronchial tumor control in patients with limited bronchogenic carcinoma. Thorax 2004;59:790-93
[21] Wang Jan T, Blackman G, George J. Survival benefits of lung cancer patients undergoing Laser an Brachytherapy . J Korean Med Sci 2002;17:341-7
[22] Escobar-Sacristán JA, Granada-Orive J, Gutierrrez T y col. Endobronchial brachytherapy in the treatment of maingant lung tumors. Eur resp J 2004;24;348-52
[23] Mehta M, Petereit D, Ghosy L y col. Sequential comparison of low-dose rate and hyperfractioned high-dose rate endobronchyal radiation for malignant airway occlutions. Int J Radiat Oncol Biol Phys 192;23(1):133-39
[24] Mallick I, Sharma S and Behera D. Endobronchyal brachytherapy for symptom palliation in non-small cell lung cancer - Analysis of symptom response, endoscopic improvement and quality of life. Lun Cancer 2007:55:313-318
[25] Bedwinek J, Petty A, Burton C y col. The use of higth dose rate endobronchyal brachytherapy to palliate symptomatic endobronchyal recurrence of previously irradiated bronchogenic carcinoma. Int J Radiat Oncol Biol Phys 1992;22:23-30
[26] Lu JJ, Bains Y, Aaron B y col. High dose rate endobronchial brachytherapy for the managent of non-small cell lung cancer with an endobronchial or peribronchial component. Cancer Therapy 2004.2:469-74
[27] Huber RM, Fischer R, Hautman H Y col. Does additional brachytherapy improve the effect of external irradiation? A prospective randomized study in central lung tumors Int J radiat Oncol Biol Phys 1997;38:53-540
[28] Mantz C, Dosoretz E, Rubeistein H y col. Endobronchial brachytherapy and optimization of local disease control in medically inoperble non-small cell lung carcinoma: A matched-pair analysis. Brachyterapy 2004;83;183.190

Section 4

Novel Strategy and Applications

6

Stereotactic Brachytherapy for Brain Tumors

Maximilian I. Ruge[1], Stefan Grau[2],
Harald Treuer[1] and Volker Sturm[1]
[1]*Department of Stereotaxy and Functional Neurosurgery,*
[2]*Department of Neurosurgery, University of Cologne,*
Germany

1. Introduction

Brachytherapy for the treatment of brain tumors has a very long history and for selected indications still represents a safe, minimally invasive and effective local treatment option, offered by specialized centers. In the following chapter we review stereotactic brachytherapy (SBT) for brain tumors, including the history, physical aspects, surgical procedure, and indications, which are introduced and discussed in the context of the available literature.

2. History of brachytherapy for brain tumors

In 1901, on Pierre Curie´s suggestion, Danlos inserted a radium isotope directly into a tumor. This is the first known use of interstitial irradiation, and was published later in 1905 in *Journal de la Physiothérapie* (Bernstein & Gutin, 1981; Danlos, 1905).

Around the same time, in 1908 Horsley and Clarke introduced a 3D targeting stereotaxis apparatus to study a monkey's brain (Horsley & Clarke, 1908). The development of stereotactic techniques had begun. The first use of this apparatus in humans was performed in 1918 by Mussen (Picard et al., 1983). This technique allowed precise targeting of brain structures for neurosurgical purposes.

The first implantation of a radioactive source into a structure of the CNS, more precisely into a tumor of the pituitary gland, was described by Hirsch in 1912 (Hirsch et al., 1912). Two years later, in 1914, Frazier reported for the first time the implantation of radioactive material into glioma (Frazier, 1920).

Further development of this technology took place in the 1930s by introducing new techniques to improve accuracy in dosimetry for multiple implants (Patterson, 1934).

In the 1950's stereotactic guidance for the implantation of radioactive sources was used to precisely treat inoperable brain tumors. (Mundinger et al., 1956; Talairach et al., 1955)

In the 1970-80s implementing image-guided stereotactic surgery continuously improved the preciseness of placing the radioactive material into (malignant) brain tumors (McDermott et al., 1998).

Since then, multiple studies have been published for the treatment of various intrinsic brain tumors, establishing (stereotactic) brachytherapy as a safe, minimally invasive, and effective neurosurgical technique for selected indications.

3. Physical aspects

Most frequently, Iodine-125-seeds (I-125) are used for stereotactic brachytherapy in brain tumors. Among a variety of possible radioactive sources used for brachytherapy, seeds containing I-125 are preferred for several reasons.

i. The energy of the emitted γ-rays (35.5 keV) and characteristic X-rays (27.2 - 31.7 keV) during the decay of I-125 is low (National Nuclear Data Center). This is advantageous, since such low energy photons are strongly absorbed in brain tissue (half-value layer: 2 cm) (Hubbell and Seltzer), yielding very high dose values inside the tumor, but additionally sparing surrounding healthy tissue. Thus the resulting dose gradient in brachytherapy with I-125-seeds is superior compared to stereotactic radiosurgery with external photon beams.
   Furthermore, radiation emitted by I-125 is strongly absorbed in cortical bone (half-value layer: 3 mm) and can easily be shielded with lead (half-value layer: 0.02 mm). Thus radiation exposure of the patient's family and health care staff can be kept low.
ii. The half-life of I-125 is rather long, 59.4 days (National Nuclear Data Center); hence the dose rate decreases by only 1.16% per day during irradiation. This allows extension of irradiation times to several weeks or longer, with only slowly decreasing dose rates. For example, a total dose of 60 Gy accumulated in 42 days equals a dose rate of as high as 1.8 Gy/day at the beginning and still 1.1 Gy/day at the end of treatment. Long-term irradiation regiments appear favorable, especially for slow-growing tumors.
iii. I-125-seeds of different type and source strength are readily available on the market (Heintz et al., 2001). The option to select seeds with activities ranging from 0.5 mCi - 10 mCi is a prerequisite for creating conformal treatment plans for brain tumors with a low number of seed-catheters. In fact, with I-125-seeds highly conformal treatment plans with conformity indices ranging from 48% - 79% (mean 70%) can be created for brain tumors by applying as few as 1 - 3 catheters (mean 1.8) and 1 - 6 seeds per catheter (mean 2.4) (Treuer et al., 2005). Implanting a low number of catheters in the brain minimizes the operative risks of bleeding and infections, and reduces operation time, and thus problems due to brain shift (Hunsche et al., 2009).

   The use of seeds with radio-opaque markers (such as Model 6711, Amersham Health) facilitates accurate intra-operative stereotactic localization of the seeds with X-rays or CT (Treuer et al., 2004). Thus the actual seed location during implantation can be determined and compared with the planned position. An accuracy of 1.5 mm in positioning of I-125-seeds was shown to be required in order to not inadequately compromise dose conformity (Treuer et al., 2005). With stereotactic guidance techniques such an accuracy requirement can be met.

   Unlike other isotopes with low quantum energy, the dosimetry of I-125-seeds appears to be well understood. The main principles of the dosimetry of interstitial brachytherapy sources were defined in the report of the Task Group 43 of the American Association of Physicists in Medicine (AAPM) in 1995 (Nath et al., 1995). Recent recommendations of the AAPM, especially for the Amersham Model 6711 I-125-seed, state that "changes in delivered dose from the introduction of the Model 6711 seed to the present have been less than 0.5% and can safely be ignored" (Willimson et al., 2005). The value of accurate and common standards in dosimetry is obvious.

4. Surgical procedure

The radioactive seed(s) can be implanted in different ways. Some authors describe the placement of seeds without stereotactic guidance ("free hand") into the tumor or the cavity after microsurgical debulking of the tumor. This procedure has been described for metastasis and gliomas (Fernandez et al., 1995; Halligan et al., 1996; Patel et al., 2000; Schulder et al., 1997; Zamorano et al., 1992). This procedure, however, makes calculation of the dose distribution in the tissue very challenging, because seed location might shift over time when the resection cavity is changing its configuration. Furthermore, a later calculation of the prescribed dose might also be challenging.

To date, the majority of groups place the seed(s) using stereotactic guidance. For this procedure a stereotactic computed tomography (CT) compatible frame is adjusted on the patient's head after inducing general or local anesthesia (Fig. 1a). We use both CT and magnetic resonance imaging (MRI) for the planning procedure: CT-imaging has the advantage of being less susceptible to distortions, while MR imaging provides better structural resolution of brain and tumor tissue. After CT imaging using a stereotactic localizer, CT scans are fused with MR images. Depending on the planning software, image fusion can be performed either automatically or by using anatomical landmarks (i.e. vessels) (Fig. 2a, CT scan; Fig 2b, MRI T1 weighed contrast). In some cases functional imaging such as functional MRI (fMRI) and/or positron emission tomography (PET) imaging can be added (C-11-methionine, F-18-FET, F-18-FDG).

The next step is defining the target volume. Using MR (T2-, T1 contrast-, and in some cases FLAIR-weighed images), CT as well as - in specific cases - PET imaging, the visible margins of the tumor are outlined manually (Fig. 2c). The objective of radiation treatment planning for brain tumors is always to determine a seed configuration with as few seed catheters as possible (to minimize operative risk) and to achieve an optimal conformation of the therapeutic isodose with respect to the surface of the target volume (Treuer et al., 2005). Inverse treatment planning is used. The desired surface dose, implantation time, and trephination point(s) are selected manually and a seed configuration yielding optimal coverage of the tumor with the prescribed dose is calculated automatically by minimization of an appropriate objective function (Bauer-Kirpes et al., 1988) (Fig. 4).

In the operation room, the stereotactic arc is adjusted on a phantom according to the calculated coordinates, and then mounted on the patient's stereotactic frame.

The ^{125}Iodine seeds (Amersham Buchler GmbH & CoKG, Braunschweig, Germany) are introduced into celcon nylon or silicon catheters (Best Industries Inc., Springfield, VA, USA / Phoenix-Biomedical Corp., Ontario, Canada; Fig. 1b, c). After skin incision and placement of an 8 mm burr hole, the catheters are inserted into the tumor (Fig. 1d). In case histology is requested, a stereotactic biopsy can be taken and evaluated during the operation. To ensure correct placement of the seed(s) intraoperatively we perform X-ray imaging in two planes (anterior/posterior and lateral) with a stationary stereotactic X-ray source, and match these images with images of the calculated trajectory (Fig. 3a, b).

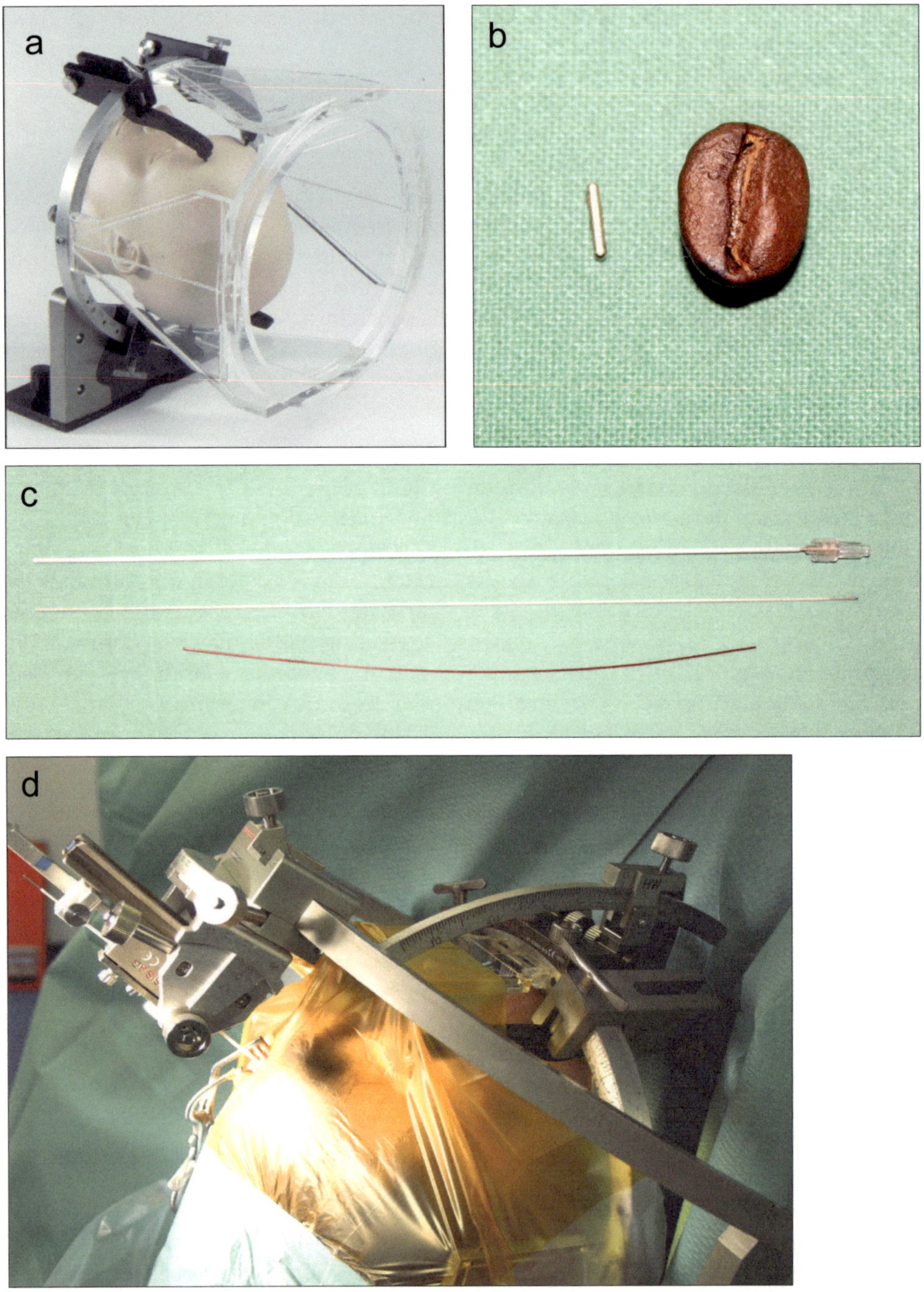

Fig. 1. Steps involved in stereotactic brachytherapy. *a) Stereotactic frame with localizer. b) Iodine-125 seed compared to a coffee bean. c) The first catheter is placed stereotactically in the target volume (above). The second catheter is filled with the calculated seed(s) (middle) which are fixed by insertion of a thin tube (below) and then placed in the first catheter. d) Operative setting with the stereotactic frame, the stereotactic arc and the inserted seed catheter.*

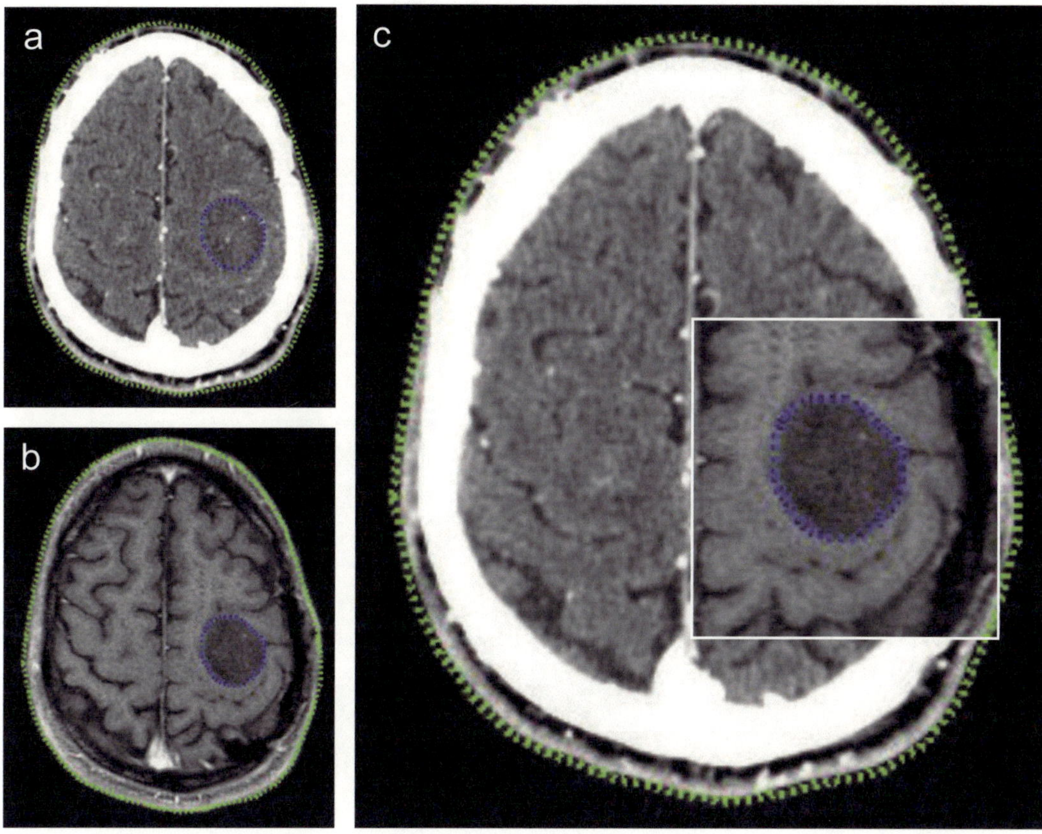

Fig. 2. Image fusion of CT and MRI scans. a) CT image; b) MRI image. c) The blue dotted line represents the manually outlined tumor margin.

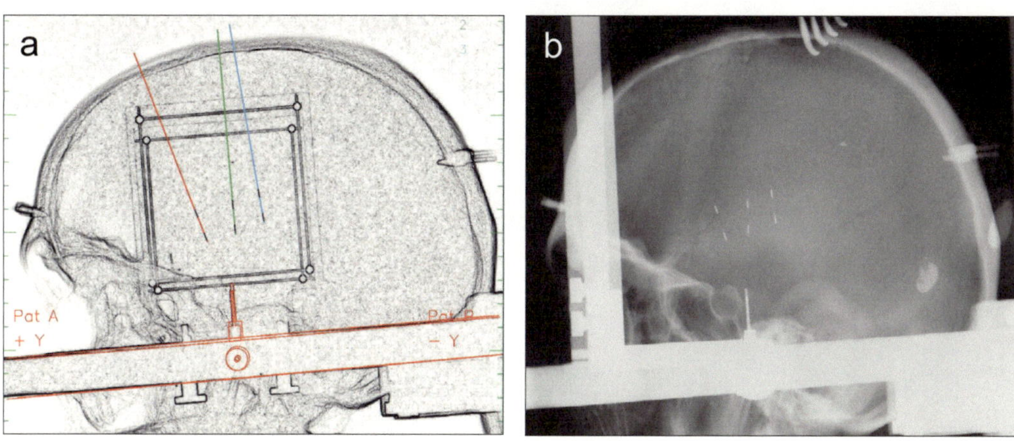

Fig. 3. The stereotactic treatment plan. a) Positioning of the catheters and seeds, which is then compared with b) the two plane X-ray images performed after final placement of the seeds-catheters by over-laying.

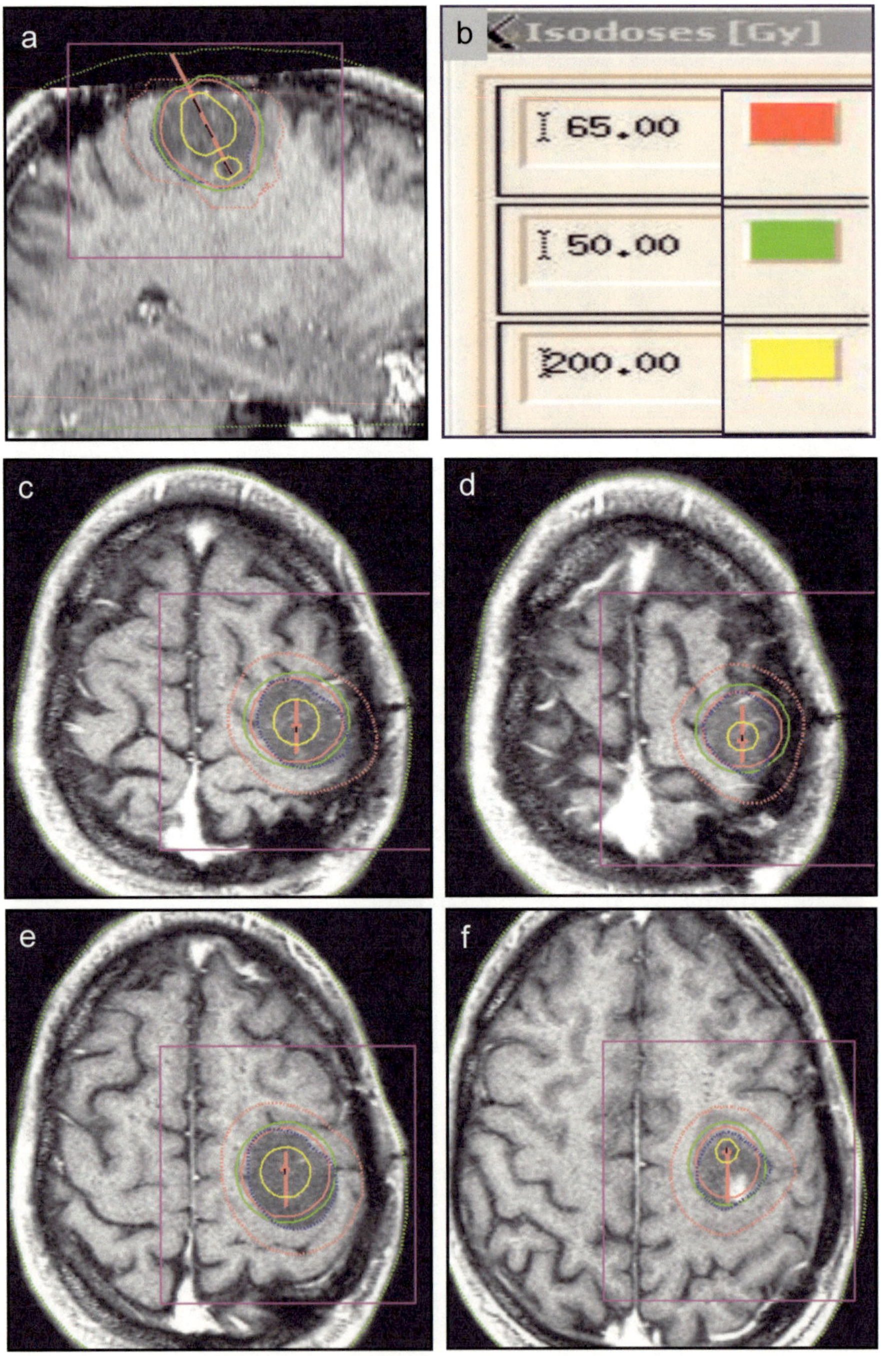

Fig. 4. An irradiation plan of an anaplastic astrocytoma WHO III in the left central sulcus area *(see also case report in Fig. 5). Isodose lines are displayed on MR T1 weighed contrast enhanced images in a) sagital, and c-f) axial orientation. The dark blue dotted line represents the manually outlined tumor surface, the green line represents the 50 Gy isodose, the red line the 65 Gy isodose and yellow the 200 Gy isodose line. b) The straight red line in a) shows the simulated catheter containing the I-125 seeds.*

Compared to intraoperative CT or MRI scanning used by some groups to monitor laser thermal therapy or placement of DBS electrodes, this technique does not require movement of the patient or a time-consuming imaging session. Furthermore, it is comparatively fast, repeatable, and even allows monitoring of the (re-)positioning of the catheters containing the I-125 seeds with high precision and with a comparatively low radiation burden for the patient (Treuer et al., 2005, Ruge et al, 2011a;b;c;d). For stabilization, the catheters are fixed within the burr hole using bone cement. In addition, the catheter tip, protruding out of the cement by approximately 3-5 mm, is fixed with a vessel clip to further avoid displacement.

In the rare cases of intraoperatively detected incorrect placement of the implanted seeds, we primarily revise the catheter position. If this revision is not satisfactory, the isodose plan is recalculated and the irradiation time adjusted accordingly.

Finally, the emitted radiation is measured at 1 m and 2 m distances from the patent's head. In cases where the dose rate exceeds 2 μSv/h at 1 m distance, the patient has to temporarily wear a lead cap (Voges et al., 1999).

The duration of the surgical procedure is usually between 40 minutes and 1.5 hours, depending on the number of catheters used. The patient's hospital stay varies between 3–5 days.

Seed catheters if implanted temporarily are removed under local anesthesia by removing the vessel clip and extracting the catheter, leaving the cement within the burr hole. This procedure requires a hospital stay of one day in most cases. At this time every patient receives a follow-up MRI using the same imaging protocol as in the SBT planning to ensure comparability.

5. Indications for Stereotactic Brachytherapy (SBT)

SBT was initially considered to be indicated for patients with a circumscribed tumor with a maximum diameter of 5 cm on CT or MRI scans. Based on available risk analysis (Kreth et al., 1997), the treatment is now widely restricted to well circumscribed tumors with a diameter not larger than 4 cm. Kreth et al. demonstrated that beyond a cut-off of approximately 3.5 cm tumor diameter (or a tumor volume of 22.4 ml) radiogenic complications increase exponentially (Kreth et al., 1997). Furthermore, the volume of the high dose irradiation zone (200 Gy isodose) also correlates directly with radiation induced tissue damage.

Patients with non-circumscribed (diffuse) tumors, or with tumors of the corpus callosum, hypothalamus, fornices, or optical system (optical nerve, chiasm) are not considered suitable for this treatment.

Among the variety of intracranial neoplasms, data from larger series exist for intracranial gliomas and metastatic brain tumors. Therefore, indication for brachytherapy should focus on these entities, while application of brachytherapy for other, rare indications, which are only mentioned briefly, should be carefully considered.

5.1 SBT for high-grade gliomas (WHO III & IV)

Malignant gliomas are non-curable intrinsic tumors of the central nervous system with a rate of almost 100% recurrence and intracranial spread. Despite improvements in different therapeutic modalities, glioblastoma still bear an overall survival rate of 12-15 months (Oertel et al., 2005; Salcman, 2001). On MRI/CT scans they usually present as a heterogeneously contrast enhancing lesion with central necrosis. With respect to treatment, radiotherapy has turned out to have the biggest impact on tumor control, time to progression, and overall survival in patients with malignant gliomas (Walker et al., 1978; 1980). To date, a multimodal approach has shown prolonged overall survival and improved quality of life (Brem et al., 1995; Stupp et al., 2005).

The extent of surgical removal of the contrast-enhancing mass has been correlated with improved overall survival (Stummer et al., 2006). However, surgical treatment is frequently limited due to the location and extension of the tumor into eloquent brain areas.

For deep-seated, non-resectable tumors, brachytherapy may theoretically be an interesting treatment alternative, affording low surgical risk and limited, local application of an effective radiation dose (Fig. 5). However, the fact that these tumors show highly invasive behavior limits this theoretically attractive approach.

Several larger studies were published predominantly in the 1980s and 1990s on brachytherapy for malignant gliomas as a primary treatment modality, or in combination with surgical removal, external beam radiation therapy (EBRT) and chemotherapy (Fermandez et al., 1995; Gutin et al., 1991; Kreth et al., 1994; Malkin et al., 1994; Scharfen et al., 1992; Sneed et al., 1998; Videtic et al., 2001; Wen et al., 1994; Zamorano et al., 1992). After thorough analysis, these studies have failed to show significant benefits in terms of time to progression (TTP) or overall survival (OS) in these patients, including patients in two randomized controlled trials (Laperriere et al., 1998; Selker et al., 2002). Furthermore, a relevant number of radiation-induced necroses requiring subsequent surgical resection were described, which is attributed to the predominantly high activity/high dose rate implantation regimens chosen due to the high proliferation rate of these tumors.

Notably, all these data were retrieved in an era prior to temozolomide, and were frequently based on CT rather than MRI findings, a fact that may further compound the derived conclusions. In the light of more recent data demonstrating the efficacy of surgery and radio-chemotherapy, brachytherapy for newly diagnosed malignant glioma does not represent a standard therapy and should not be favored over standard therapy. There is also no recent evidence supporting a combined approach with brachytherapy after surgical resection and in combination with EBRT.

Novel approaches combining brachytherapy with EBRT and chemotherapy for deep-seated, non-respectable tumors as a first line treatment are under clinical investigation, but data are not available yet.

As mentioned above, malignant gliomas show extremely high recurrence rates despite intensive and combined therapy. Therefore, the first studies on brachytherapy for intracranial tumors were performed on recurrent gliomas. In this particular setting, a series of permanent or temporary implants with different dose rates were reported to have variable outcomes in terms of overall survival and progression free survival (Gutin et al., 1984; 1987; Halligan et al., 1996; Julow et al., 2007; Kitchen et al., 1994; Larson et al., 2004; Patel et al., 2000; Ryken et al., 1994; Tselis et al., 2007). The published studies showed no evidence for a significant benefit from brachytherapy in recurrent glioblastoma, although probably a subset of patients may benefit from this treatment. Nevertheless, the method still may be applied as a salvage option in selected patients.

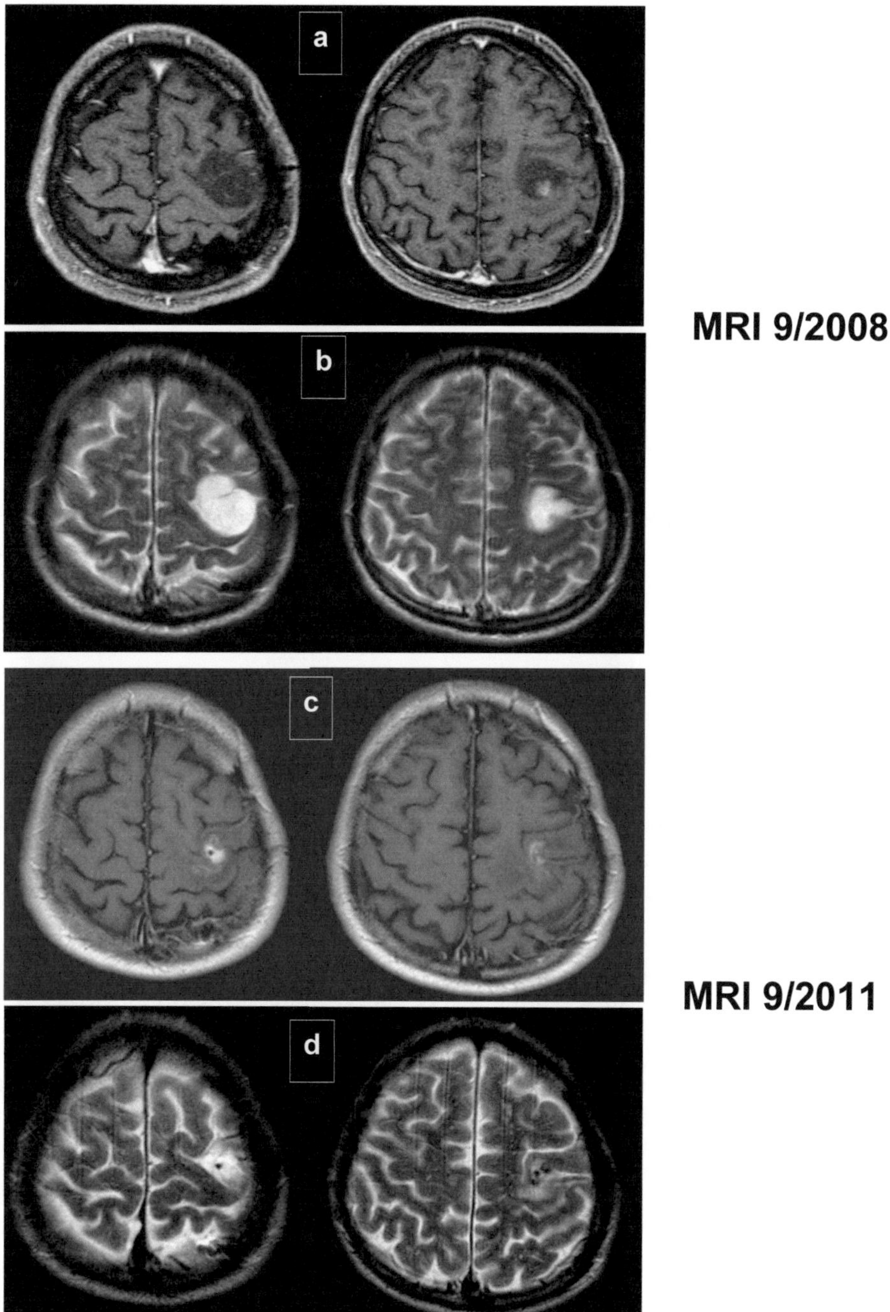

Fig. 5. MRI of a patient with anaplastic astrocytoma WHO III. *MRI scans of an 41 year old patient diagnosed by stereotactic biopsy with an anaplastic astrocytoma WHO III in September 2008; a) T1- with contrast, b) T2-weighed images. According to our protocol he underwent stereotactic brachytherapy (I-125, 50 Gy cumulative tumor surface dose, permanent implantation) followed by 25 Gy EBRT boost. c, d) Follow-up MRI shows partial response of the tumor two years later. The patient's initial paresis of the right hand resolved completely.*

5.2 SBT for low-grade gliomas (WHO I & II)

Considerable data exist for brachytherapeutic treatment of low-grade gliomas (WHO I pilocytic astrocytomas, WHO II astrocytomas, oligoastrocystomas and oligodendrogliomas), both in children and adults (Julow et al., 2007; Korinthenberg et al., 2010; Kreth et al., 1995; 2006; Mehrkens et al., 2004; Mundinger et al., 1991; Ostertag et al., 1993; Ruge et al., 2011a; Schatz et al., 1994; Suchorska et al., 2011).

The biology if these tumors diverges from malignant gliomas, with a much lower proliferation index and cellular density. They may occur in a diffuse manner, but are frequently diagnosed as rather well delineated, circumscribed lesions usually lacking contrast enhancement. For the latter form, brachytherapy has been applied in a large number of patients.

The role of surgery for these tumor remains to be clarified, although the tendency is a positive impact of gross total resection upon overall survival and time to progression (Frappaz et al., 2002; Nunez et al., 2009; Sanai et al., 2008).

The theoretical background for brachytherapy in low-grade gliomas is different from that in malignant gliomas. In accordance with the low proliferation index and presumed co-existence of still functioning healthy brain tissue within the tumor, most stereotactic neurosurgeons used low activity implants, either temporary or permanent, which also resulted in very low dose rates.

The reported 5- and 10-year survival rates ranged from 61% and 51% for grade II (with the exception of gemistocytic gliomas) up to 85% and 83% for pilocytic gliomas (Kreth et al., 1995; Ruge et al, 2011a; Suchorska et al, 2011). These data represent an at least equal outcome compared to other frequently used therapies such as surgery and EBRT (Rock, 1999). Furthermore, procedure-related risk is low (<5%), and long and short-term complications of brachytherapy using low dose rate implants are lower than those associated with tumor resection and/or EBRT (Kreth et al., 1995; Sarkaria et al., 1995; Shrieve et al., 1995).

Important to point out is the fact that SBT can be successfully combined with up-front neurosurgical partial resection of safely accessible portions of the tumor (Ruge et al., 2011a; Schnell et al., 2008; Suchorska et al., 2011).

As a consequence, according to data, brachytherapy may represent an adequate treatment option for WHO I & II gliomas. However, all these data were retrieved from retrospective analyses and large, controlled prospective trials are still lacking.

Pilocytic astrocytomas (WHO I) are the most benign lesions among intrinsic brain tumors. They most frequently occur in children and may be cured by total surgical removal. The underlying reason for this is their limited invasion into surrounding nervous tissue, which makes even histologically complete removal possible. Therefore in these tumors surgery is still considered the treatment of choice in cases where complete removal can be performed without impairment for the patient (Burkhard et al., 2003; Kayama, 1996; Morreale et al., 1997). As might be expected, incomplete resection reduced overall survival and time to progression significantly (Morreale et al., 1997).

However, pilocytic astrocytomas frequently occur in deep-seated locations adjacent to vital nervous structures, which can make removal of these tumors extremely dangerous or even impossible. In such cases, brachytherapy may be an effective and safe treatment option, especially due to the histological setting mentioned above. Two recent studies of a large series of children treated with SBT for pilocytic astrocytoma reported five-year progression free survival of 91% and 92% (Korinthenberg et al., 2011; Ruge et al., 2011a). Treatment-related complications were as low as 5.4% (Ruge et al, 2011a). For extensive pilocytic astrocytomas, combined treatment approaches using surgery and brachytherapy have been used and are under clinical investigation (Peraud et al., 2007; Ruge et al., 2011a).

Again, it should be mentioned that these data concerning grade I gliomas are also retrospective, and there is no collected evidence based on prospective, controlled trials.

In summary, to date there is no evidence that patients with malignant gliomas benefit from brachytherapy, and this treatment should not be considered a first line option. However, all data were retrieved in an era before temozolomide, and molecular markers were available. Therefore, a subset of patients may still benefit from brachytherapy, and they need to be defined in future trials.

To a similar extent, data on low-grade gliomas have to be interpreted carefully, since no prospective controlled trials have been carried out, and data derives exclusively from retrospective studies. Nevertheless, time to progression and overall survival were equal or even superior to other treatment modalities, with a lower level of treatment-related morbidity and mortality, and shortened hospital stay. Furthermore, the use of SBT does not limit the use of EBRT in cases of tumor progression or relapse.

5.3 SBT for metastatic disease

Apart from intrinsic brain tumors, brachytherapy has been used as an adjuvant treatment modality for metastatic brain tumors. Although there are more, high level data concerning the primary treatment of brain metastasis (i.e. EBRT, radiosurgery, surgery), brachytherapy may be a treatment option for recurrent or *de novo* metastasis that cannot be treated by evidence-based conventional modalities.

In these cases it has been shown that brachytherapy may lead to good local tumor control with low morbidity and no treatment related mortality. Furthermore, due to the low surgical burden the procedure may even be performed on patients in a generally reduced condition as a salvage treatment (Bogart et al., 1999; Kreth et al., 1995; Ostertag et al., 1995; Ruge et al, 2011c). Furthermore, SBT allows histological diagnosis within the same surgical procedure, which is useful in newly occurring cerebral lesions without a proven systemic tumor, so-called CUP syndrome (CUP, cancer of unknown primary) or in cases of new or progressive local tissue changes after stereotactic radiosugery (Ruge et al., 2011b)

In summary, defining a selection of patients who may benefit from brachytherapy will remain one of the most challenging tasks. In this context, brachytherapy may represent a useful tool in children and adults with deep-seated and non-resectionable WHO I and II tumors, as well as in patients with cerebral metastatic tumors in limited numbers, which are not amenable to other therapies (salvage treatment).

Only for low-grade gliomas WHO I and II, as well as metastatic brain tumors is there substantial evidence for a beneficial effect in terms of time to progression and overall survival. However, also for these populations lager, prospective, randomized and controlled trials are still lacking, thus available data remain limited in their message due to their retrospective nature.

5.4 Rare indications

The use of brachytherapy has been reported in atypical meningeomas, pineal gland tumors and medulloblastomas (Maarouf et al., 2010; El Majdoub et al., 2011; von Hoff et al., 2009; Ware et al., 2004).

For meningeomas, besides surgical removal, radio-surgery has gained considerable importance over the past few decades. In the light of technical advances in radiosurgical devices and more widespread availability, brachytherapy does not play a role in the management of intracranial meningeomas.

For medulloblastoma, a highly malignant tumor entity that primarily occurs in children, brachytherapy may only have its place as a salvage option due to the aggressively spreading nature of these tumors. There are no data showing any superiority of brachytherapy over evidence-based approaches such as surgery followed by radiation and chemotherapy.

6. Conclusion

SBT represents a safe, minimally invasive and highly effective local treatment concept always aiming for the treatment of the entire tumor. It can be repeated and does not hinder or limit the use of EBRT in cases of further tumor progression. The possibility of combining SBT with partial tumor resection represents a feasible concept that has already been described in a subsets of patients. To further establish SBT as a ubiquitously available local treatment concept equal to open microneurosurgery or stereotactic radiosurgery, prospective randomized trials, especially for low-grade gliomas, are warranted.

7. References

Bauer-Kirpes B, Sturm V, Schlegel W, Lorenz WJ. Computerized optimization of 125I implants in brain tumours. Int J Radiat Oncol Biol Phys. 1988;14:1013–23.

Bernstein M, Gutin PH. Interstitial irradiation of brain tumors: a review. Neurosurgery. 1981;9(6):741-50.

Bogart JA, Ungureanu C, Shihadeh E et al. Resection and permanent I-125 brachytherapy without whole brain irradiation for solitary brain metastasis from non-small cell lung carcinoma. J Neurooncol. 1999;44(1): 53-57.

Brem H, Piantadosi S, Burger PC, Walker M, Selker R, Vick NA, Black K, Sisti M, Brem S, Mohr G; Placebo-controlled trial of safety and efficacy of intraoperative controlled delivery by biodegradable polymers of chemotherapy for recurrent gliomas. The Polymer Brain Tumor Treatment Group. Lancet. 1995;345:1008–1012.

Brem H, Ewend MG, Piantadosi S, Greenhoot J, Burger PC, Sisti M. The safety of interstitial chemotherapy with BCNU-loaded polymer followed by radiation therapy in the treatment of newly diagnosed malignant gliomas: phase I trial. J Neurooncol. 1995;26:111–123.

Burkhard C, Di Patre PL, Schüler D, Schüler G, Yasargil G, Yonekawa Y, Lütolf UM, Kleihues P, Ohgaki H. A population-based study of incidence and survival rates in patients with pilocytic astrocytoma. J Neurosurg. 2003;98:1170-1174.

Danlos H. Quelques considerations sur le traitement des dermatoses par le radium. J Physiotherapie (Paris). 1905;3: 98-106.

El Majdoub F, Simon T, Hoevels M, Berthold F, Sturm V, Maarouf M. Interstitial brachytherapy using stereotactic implanted (125)iodine seeds for recurrent medulloblastoma. Clin Oncol (R Coll Radiol). 2011;23(8):532-7.

Fernandez P, Zamorano L, Yakar D, Gaspar L, Warmelink C. Permanent Iodine-125 implants in the up-front treatment of malignant gliomas. Neurosurgery. 1995;36: 467-73.

Frappaz D, Chinot O, Bataillard A, Ben Hassel M, Capelle L, Chanalet S, Chatel M, Figarella-Branger D, Guegan Y, Guyotat J, Hoang-Xuan K, Jouanneau E, Keime-Guibert F, Laforet C, Linassier C, Loiseau H, Maire JP, Menei P, Rousmans S, Sanson M, Sunyach MP. Summary version of the Standards, Options and Recommendations for the management of adult patients with intracranial glioma. Br J Cancer. 2002;89: 73–83.

Frazier CH The effects of radium emanations upon brain tumors. Surg Gyn Obstet. 1920;31: 236-239.

Gutin PH, Prados MD, Phillips TL, Wara WM, Larson DA, Leibel SA, Sneed PK, Levin VA, Weaver KA, Silver P, et al.External irradiation followed by an interstitial high activity iodine-125 implant "boost" in the initial treatment of malignant gliomas: NCOG study 6G-82-2. Int J Radiat Oncol Biol Phys. 1991;21(3):601-6.

Gutin PH, Leibel SA, Wara WM, Choucair A, Levin VA, Philips TL, Silver P, Da Silva V, Edwards MS, Davis RL, et al.Recurrent malignant gliomas: survival following interstitial brachytherapy with high-activity iodine-125 sources.J Neurosurg. 1987;67(6):864-73. Review.

Halligan JB, Stelzer KJ, Rostomily RC, Spence AM, Griffin TW, Berger MS. Operation and permanent low activity 125I brachytherapy for recurrent high-grade astrocytomas. Int J Radiat Oncol Biol Phys. 1996;35: 541-547.

Heintz B H, Wallace R E, Hevezi J M. Comparison of I-125 sources used for permanent interstitial implants. Med. Phys. 2001;28: 671-82.

Hirsch O. Die operative Behandlung von Hypophysentumoren: Nach endonasalen Methoden. Arch Laryngol Rhinol. 1912;26:529-686.

Horsley V, Clarke RH. On a method of investigating the deep ganglia and tracts of the central nervous system (cerebellum). Br Med J. 1906;1799-1800.

Hubbell J H, Seltzer S M. Tables of X-Ray Mass Attenuation Coefficients and Mass Energy-Absorption Coefficients from 1 keV to 20 MeV for Elements Z = 1 to 92 and 48 Additional Substances of Dosimetric Interest. http://www.nist.gov/pml/data/xraycoef/index.cfm, accessed 08-09-2011

Hunsche S, Sauner D, Maarouf M, Poggenborg J, Lackner K, Sturm V, Treuer H. Intraoperative X-ray detection and MRI-based quantification of brain shift effects subsequent to implantation of the first electrode in bilateral implantation of deep brain stimulation electrodes. Stereotact. Funct. Neurosurg. 2009; 87: 322-9.

Julow J, Viola A, Bálint K, Szeifert GT. Image fusion-guided stereotactic iodine-125 interstitial irradiation of inoperable and recurrent gliomas. Prog Neurol Surg. 2007;20:303-11.

Kayama T, Tominaga T, Yoshimoto T. Management of pilocytic astrocytoma. Neurosurg Rev. 1996;19: 217-220.

Kitchen ND, Hughes SW, Taub NA, Sofat A, Beaney RP, Thomas DG.Survival following interstitial brachytherapy for recurrent malignant glioma. J Neurooncol. 1994;18(1):33-9.

Korinthenberg R, Neuburger D, Trippel M, Ostertag C, Nikkhah G. Long-term results of brachytherapy with temporary iodine-125 seeds in children with low-grade gliomas. Int J Radiat Oncol Biol Phys. 2011;79: 1131-8.

Kreth FW, Warnke PC, Ostertag CB. Long-term follow-up results of 175 patients with malignant glioma. Acta Neurochir (Wien). 1994;131(1-2):164-6.

Kreth FW, Faist M, Warnke PC, Rossner R, Volk B, Ostertag CB. Interstitial radiosurgery of low-grade gliomas. J Neurosurg. 1995;82: 418-29.

Kreth FW, Faist M, Rossner R, Birg W, Volk B, Ostertag CB The risk of interstitial radiotherapy of low-grade gliomas. Radiother Oncol. 1997;43(3):253-60.

Kreth FW, Faist M, Grau S, Ostertag CB. Interstitial 125 radiosurgery of supratentorial de novo WHO Grade 2 astrocytoma and oligoastrocytoma in adults: long-term results and prognostic factors. Cancer. 2006;106:1372-81.

Laperriere NJ, Leung PMK, McKenzie S, Miloevic M, Wong S, Glen J, Pintilie M, Bernstein M: Randomized study of brachytherapy in the initial management of patients with malignant astrocytoma. Int J Radiat Oncol Biol Phys. 1998;41: 1005–1011.

Larson DA, Suplica JM, Chang SM, Lamborn KR, McDermott MW, Sneed PK, Prados MD, Wara WM, Nicholas MK, Berger MS.Permanent iodine 125 brachytherapy in patients with progressive or recurrent glioblastoma multiforme. Neuro Oncol. 2004;6(2):119-26.

Leksell, L. A stereotaxic apparatus for intracerebral surgery.Acta Chir Scand. 1949;99: 229-233.

Maarouf M, El Majdoub F, Bührle C, Voges J, Lehrke R, Kocher M, Hunsche S, Treuer H, Sturm V. Pineal parenchymal tumors. Management with interstitial iodine-125 radiosurgery. Strahlenther Onkol. 2010;186(3):127-34.

Malkin MG. Interstitial brachytherapy of malignant gliomas: the Memorial Sloan-Kettering Cancer Center experience. Recent Results Cancer Res. 1994;135: 117-25.

McDermott MW, Sneed PK, Gutin PH.Interstitial brachytherapy for malignant brain tumors. Semin Surg Oncol. 1998;14(1):79-87.

Morreale VM, Ebersold MJ, Quast LM, et al. Cerebellar astrocytoma: experience with 54 cases surgically treated at the Majo Clinic, Rochester, Minnesota, from 1978 to 1990. J Neurosurg. 1997;87: 257-261.

Mehrkens JH, Kreth FW, Muacevic A, Ostertag CB.Long term course of WHO grade II astrocytomas of the Insula of Reil after I-125interstitial irradiation. J Neurol. 2004;251(12):1455-64.

Mundinger, F. Eine einfache Methode der lokalisierten Bestrahlung von Großhirngschwülsten mit radioaktivem Gold. Münch Med Wschr. 1956;98:23-25.

Mundinger, F (1958) Beitrag zur Dosimetrie und Applikation von Radio-Tanal (182Ta) zur Langzeitbestrahlung von Hirngeschwülsten. Fortsch. Röntgenstr. 1958;89: 86-91.

Mundinger F, Braus DF, Krauss JK, Birg W. Long-term outcome of 89 low-grade brain-stem gliomas after interstitial radiation therapy. J Neurosurg. 1991;75:740-6.

Narabayashi H, Okuma R. Procain oil blocking of the globus pallidus for the treatment of rigidity and tremor of parkinsonisma (preliminary report). Proc Jap Acad. 1953;29: 134-137.

National Nuclear Data Center, Brookhaven National Laboratory. http://www.nndc.bnl.gov/chart/, accessed 08-09-2011.

Nath R, Anderson L L, Luxton G, Weaver K A, Williamson J F, Meigooni A S. Dosimetry of interstitial brachytherapy sources: recommendations of the AAPM Radiation Therapy Committee Task Group No. 43. Med. Phys. 1995;22:209-36.

Nunez OM, Seol HJ, Rutka JT. The role of surgery in the management of intracranial gliomas: current concepts. Indian J Cancer. 2009;46(2):120-6.

Oertel J, von Buttlar E, Schroeder HWS, Gaab MR Prognosis of gliomas in the 1970s and today. Neurosurg Focus, 2005;18:12.

Ostertag CB.Brachytherapy--interstitial implant radiosurgery.Acta Neurochir Suppl (Wien). 1993;58:79-84.

Ostertag CB, Kreth FW. Interstitial iodine-125 radiosurgery for cerebral metastases. Br J Neurosurg. 1995;9:593-603.

Patel S, Breneman JC, Warnick RE, Albright RE Jr, Tobler WD, van Loveren HR, Tew JM Jr. Permanent iodine-125 interstitial implants for the treatment of recurrent glioblastoma multiforme. Neurosurgery. 2000;46: 1123-8.

Patterson R. A dosage system for gamma ray therapy: Part I. Br J Radiol 1934;7:592-612

Peraud A, Goetz C, Siefert A, Tonn JC, Kreth FW. Interstitial iodine-125 radiosurgery alone or in combination with microsurgery for pediatric patients with eloquently located low-grade glioma: a pilot study. Childs Nerv Syst. 2007;23(1):39-46. Epub 2006 Sep 14.

Picard C, Olivier A, Bertrand G. The first human stereotaxic apparatus. The contribution of Aubrey Mussen to the field of stereotaxis. J Neurosurg. 1983;59(4):673-6.

Riechert T, Mundinger F. Beschreibung und Anwendung eines Zielgerätes für stereotaktische Hirnoperationen. Acta Neurochir (Wien). 1955;3: 308-337.

Ruge MI, Simon T, Suchorska B, Lehrke R, Hamisch C, Koerber F, Maarouf M, Treuer H, Berthold F, Sturm V, Voges J. Stereotactic Brachytherapy With Iodine-125 Seeds for the Treatment of Inoperable Low-Grade Gliomas in Children: Long-Term Outcome. J Clin Oncol. 2011a.

Ruge MI, Kickingereder P, Grau S, Hoevels M, Treuer H, Sturm V. Stereotactic biopsy combined with stereotactic (125)iodine brachytherapy for diagnosis and treatment of locally recurrent single brain metastases. J Neurooncol. 2011b;105:109-18.

Ruge MI, Suchorska B, Maarouf M, Runge M, Treuer H, Voges J, Sturm V. Stereotactic 125iodine brachytherapy for the treatment of singular brain metastases: closing a gap? Neurosurgery. 2011c;68:1209-18.

Ruge MI, Kocher M, Maarouf M, Hamisch C, Treuer H, Voges J, Sturm V. Comparison of stereotactic brachytherapy (125 iodine seeds) with stereotactic radiosurgery (LINAC) for the treatment of singular cerebral metastases. Strahlenther Onkol. 2011d;187:7-14.

Ryken TC, Hitchon PW, VanGilder JC, Wen BC, Jani S.Interstitial brachytherapy versus cytoreductive surgery in recurrent malignant glioma. Stereotact Funct Neurosurg. 1994;63(1-4):241-5.

Salcman, M, Glioblastoma multiforme and anaplastic astrocytoma, In: Kaye AH, Laws ER Jr, ed Brain Tumors Londen Churchill Livingstone, 2001;493-523.

Sanai N, Berger MS. Glioma extent of resection and its impact on patient outcome. Neurosurgery. 2008;62:753-64.

Sarkaria JN, Mehta MP, Loeffler JS, Buatti JM, Chappell RJ, Levin AB, Alexander E 3rd, Friedman WA, Kinsella TJ. Radiosurgery in the initial management of malignant gliomas: survival comparison with the RTOG recursive partitioning analysis. Radiation Therapy Oncology Group. Int J Radiat Oncol Biol Phys. 1995;32(4):931-41.

Scharfen CO, Sneed PK, Wara WM, Larson DA, Phillips TL, Prados MD, Weaver KA, Malec M, Acord P, Lamborn KR et al. High activity iodine-125 interstitial implant for gliomas. Int J Radiat Oncol Biol Phys. 1992;24:583–591.

Schätz CR, Kreth FW, Faist M, Warnke PC, Volk B, Ostertag CB. Interstitial 125-iodine radiosurgery of low-grade gliomas of the insula of Reil. Acta Neurochir (Wien). 1994;130:80-9.

Selker RG, Shapiro WR, Burger P, Blackwood MS, Arena VC, Gilder JC, Malkin MG, Mealey JJ Jr, Neal JH, Olson J, Robertson JT, Barnett GH, BloomWeld S, Albright R, Hochberg FH, Hiesiger E, Green S. The Brain Tumor Cooperative Group NIH Trial 87–01: a randomized comparison of surgery, external radiotherapy, and carmustine versus surgery, interstitial radiotherapy boost, external radiation therapy, and carmustine. Neurosurgery. 2002;51:343–355.

Shrieve DC, Loeffler JS. Advances in radiation therapy for brain tumors. Neurol Clin. 1995;13(4):773-93. Review.

Schnell O, Schöller K, Ruge M, Siefert A, Tonn JC, Kreth FW. Surgical resection plus stereotactic 125I brachytherapy in adult patients with eloquently located supratentorial WHO grade II glioma - feasibility and outcome of a combined local treatment concept. J Neurol. 2008;255(10):1495-502.

Schulder M, Black PM, Shrieve DC, Alexander E 3rd, Loeffler JS. Permanent low-activity iodine-125 implants for cerebral metastases. J Neurooncol. 1997, 33: 213-21. Schurr PH, Merrington WR. The Horsley-Clarke stereotaxic apparatus. Br J Surg. 1978;65(1):33-6.

Spiegel EA, Wycis HT, Marks M, Lee AJ. Stereotaxic apparatus for operations on the human brain. Science. 1947;106: 349-350.

Sneed PK, StauVer PR, McDermott MW, Diederich CJ, Lamborn KR, Prados MD, Chang S, Weaver KA, Spry L, Malec MK, Lamb SA, Voss B, Davis RL, Wara WM, Larson DA, Phillips TL, Gutin PH. Survival benefit of hyperthermia in a prospective randomized trial of brachytherapy boost § hyperthermia for glioblastoma multiforme. Int J Radiat Oncol Biol Phys. 1998;40:287–295.

Stummer W, Pichlmeier U, Meinel T, et al. Fluorescence-guided surgery with 5-aminolevulinic acid for resection of malignant glioma: a randomised controlled multicentre phase III trial. Lancet Oncol. 2006;7:392–401.

Stupp R, Mason WP, van den Bent MJ, Weller M, Fisher B, Taphoorn MJ, et al. Radiotherapy plus concomitant and adjuvant temozolomide for glioblastoma. N Engl J Med. 2005;352:987-996.

Suchorska B, Ruge M, Treuer H, Sturm V, Voges J. Stereotactic brachytherapy of low-grade cerebral glioma after tumor resection. Neuro Oncol. 2011;13:1133-42. Epub 2011 Aug 25.

Talairach J, Ruggiero G, Aboulker, J, David M. A method of treatment of inoperable brain tumours by stereotaxic implantation of radioactive gold. Br J Radiol. 1955;28: 62-74.

Tselis N, Kolotas C, Birn G, Roddiger S, Filipowicz I, Kontova M, Fountzilas G, Selviaridis P, Baltas D, Heyd R, Anagnostopoulos G, Zamboglou N (2007) CT-guided interstitial HDR brachytherapy for recurrent glioblastoma multiforme. Long-term results. Strahlenther Onkol. 2007;183:563–570.

Treuer H, Hunsche S, Hoevels M, Luyken K, Maarouf M, Voges J, V Sturme. The influence of head frame distortions on stereotactic localization and targeting. Phys. Med. Biol. 2004; 49: 3877-87.

Treuer H, Klein D, Maarouf M, Lehrke R, Voges J, Sturm V. Accuracy and conformity of stereotactically guided interstitial brain tumor therapy using I-125 seeds. Radiother. Oncol. 2005; 77(2):202-9.

Videtic GM, Gaspar LE, Zamorano L, et al.: Implant volume as a prognostic variable in brachytherapy decisionmaking for malignant gliomas stratified by the RTOG recursive partitioning analysis. Int J Radiat Oncol Biol Phys. 2001;51:963–968.

Voges J, Sturm V. Interstitial irradiation with stereotactically implanted I-125 seeds for the treatment of cerebral glioma. Crit Rev Neurosurg. 1999;9(4):223–233.

von Hoff K, Hinkes B, Gerber NU, Deinlein F, Mittler U, Urban C, Benesch M, Warmuth-Metz M, Soerensen N, Zwiener I, Goette H, Schlegel PG, Pietsch T, Kortmann RD, Kuehl J, Rutkowski S. Long-term outcome and clinical prognostic factors in children with medulloblastoma treated in the prospective randomised multicentre trial HIT'91.Eur J Cancer. 2009;45(7):1209-17. Epub 2009 Feb 26.

Walker MD, Alexander E Jr, Hunt WE, MacCarty CS, Mahaley MS Jr, Mealey J Jr, Norrell HA, Owens G, Ransohoff J, Wilson CB, Gehan EA, Strike TA. Evaluation of BCNU and/or radiotherapy in the treatment of anaplastic gliomas. A cooperative clinical trial. J Neurosurg. 1978;49(3):333-43.

Walker MD, Green SB, Byar DP, Alexander E Jr, Batzdorf U, Brooks WH, et al. Randomized comparisons of radiotherapy and nitrosoureas for the treatment of malignant glioma after surgery. N Engl J Med. 1980;303:1323-1329.

Ware ML, Larson DA, Sneed PK, Wara WW, McDermott MW. Surgical resection and permanent brachytherapy for recurrent atypical and malignant meningioma. Neurosurgery. 2004;54(1):55-63.

Wen PY, Alexander E 3rd, Black PM, Fine HA, Riese N, Levin JM, Coleman CN, Loeffler JS: Long term results of stereotactic brachytherapy used in the initial treatment of malignant gliomblastoma. Cancer. 1994;73: 3029–3036.

Williamson J F, Butler W, DeWerd L A, Saiful Huq M, Ibbott G S, Li Z, Mitch M G, Nath R, Rivard M J, Todor D. Recommendations of the American Association of Physicists in Medicine regarding the Impact of Implementing the 2004 Task Group 43 Report on Dose Specification for 103Pd and 125I Interstitial Brachytherapy. Med Phys. 2005;32:1424-39.

Zamorano L, Yakar D, Dujovny M, Sheehan M, Kim J. Permanent iodine-125 implant and external beam radiation therapy for the treatment of malignant brain tumors. Stereotact Funct Neurosurg. 1992;59:183-92.

Safe and Curative Brachytherapy Reirradiation with Organ-Sparing Hyaluronate Gel Injection

Kazushi Kishi, Yasutaka Noda and Morio Sato
Wakayama Medical University,
Japan

1. Introduction

In radiotherapy, sufficient dose delivery promises local cure [Hayashi, 2002; Okamoto, 2002; Wu, 2007] and reirradiation is as effective as the first radiotherapy. Dose dependency of the effect is also reported in reirradiation [Damast, 2010]. In the clinical situation, however, the use of reirradiation is limited because of the low tolerance level of surrounding normal tissue [Emami, 1991]. A significant incidence of late toxicity attributable to accumulated dose in various at-risk organs is reported [Salama, 2006; Wu, 2007]. In a survey of the attitudes of radiation oncologists, only one third (90 of 271) of participants responded that they would consider reirradiation for in-field failure after previous radical radiotherapy [Joseph, 2008].

Recent advancements in external beam radiation technology have enabled precision by beam set-up in image-guided radiotherapy (IGRT), accuracy in configuring dose distribution by intensity modified radiotherapy (IMRT), and adaptability to internal movements by real-time tracking techniques. However, surgery has been the only option for saving at-risk organs closely attached to the target. It is desirable to establish non-invasive or minimally invasive techniques for preserving these at-risk organs to enable safe and curative high-dose radiotherapy, even in the case of reirradiation.

- For the purpose of reirradiation, we devised a minimally invasive procedure using a modern high-molecular-weight hyaluronate polymer gel, whereby a planned distance is created between the target and at-risk organs by injecting the gel (Fig. 1) [Kishi, 2004 、2006, 2007a, 2007b, 2009, Prada, 2007, 2009; Vordermark, 2008].

Fig. 1. An organ at risk (OAR) is displaced away from the zone of intensive irradiation (dotted circle) by the space created by an injected matrix.

In the first stage of our strategy, we aimed to treat recurrent disease requiring reirradiation with single-session high-dose-rate brachytherapy (HDRBT), reporting an effective decrease in estimated cumulative normal tissue complication probabilities (NTCP) to safer levels. In the second stage, we established this procedure for various difficult situations such as reirradiation of para-aortic lymphnode metastasis.

2. Materials and methods

2.1. Native-type hyaluronate as an injectable spacer

2.1.1 Materials for injectable spacer

Historically, saline, autologous blood, olive oil, dextrose, hyaluronate (hyaluronic acid: HA), alginate gels, gelatin hydrogel, polytetrafluoroethylene sheets, and many other others materials have been used for the purpose of separating various tissues. We have used injectable spacers for risk-organ preservation during brachytherapy since 2004 [Kishi, 2004, 2007, 2007b]. Gel dissection has the features of both blunt dissection and hydrodissection, depending on its viscosity and the injection rate. Important considerations in selecting an injectable spacer include its chemical and biological inertness or activity, allergenicity, and the natural or artificial biodegradability of the material; the structure of the injected space is also an important factor, which is discussed later.

2.1.2 Native-type hyaluronate molecules

The numerous roles of innate hyaluronate in the body include as a supportive component of structure, lubrication in joints, tissue protection, and wound healing. Hyaluronate (also called hyaluronic acid or hyaluronan) is an anionic, nonsulfated macromolecular glycosaminoglycan, composed of repeated disaccharide units of glucuronic acid and N-acetylglucosamine (Fig. 2), with a widely variable molecular weight. The native type is a single straight chain and exists widely, mainly as a constitutive molecule in the extracellular space of human and animal tissue. The molecular weight of constitutive hyaluronate is about three million Daltons. Degradation of hyaluronate occurs during inflammation and in the injury healing process, and also during bacterial infection. of innate or bacterial hyaluronidase to lower-molecular-weight hyaluronate and in further biodegradation. The degraded hyaluronate promotes tissue reactions.

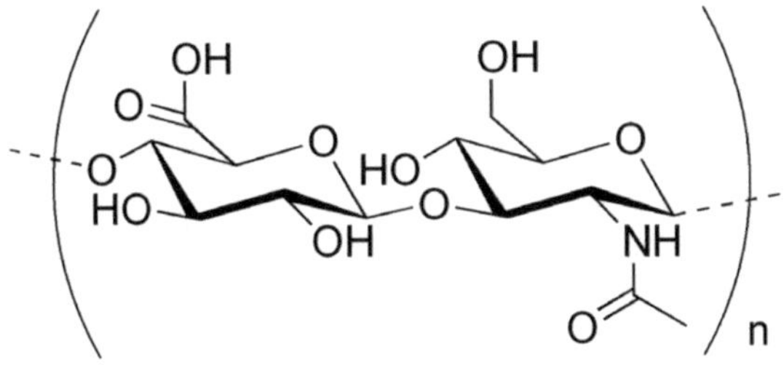

Fig. 2. Molecular structure of hyaluronate

The known cellular surface receptors CD44, RHAMM (receptor for hyaluronic-acid-mediated motility), and ICAM-1 (intercellular adhesion molecule-1) have hyaluronate binding sites that regulate cellular migration, proliferation, and inflammatory responses.

The binding activity is partially regulated by the molecular size of the hyaluronate: high-molecular-weight hyaluronate (HMW-HA) acts to almost completely inhibit the above responses, while low-molecular-weight hyaluronate has a promotional effect. Recent reports state that HMW-HA inhibits the production of matrix metalloproteinases [Hashizume, 2010] as well as vascular leakiness [Singleton, 2010], and that there is a size function of hyaluronate [Wolny. 2010]. It may be speculated that a partially combined huge mass made of ligand masks itself until degradation to a size small enough to interact diffusely with the receptors, and which is constitutively maintained in normal tissue. In contrast, artificially cross-linked hyaluronate may have no or various features of this inhibition–promotion mechanism.

In medical use, the effectiveness of native-type hyaluronate has been previously reported in the management of radiation dermatitis [41] [Primavera, 2006], in the prevention of postoperative adhesions in the pelvis [Kusunoki, 2005], and as antioxidant [Campo, 2008], depending on its molecular weight and concentration [Krasinski, 2009].

A high-molecular-weight hyaluronate that is commercially available for articular space injection (3.4 million Daltons of median molecular weight, 2.2 million of viscosity molecular size; Suvenyl, Chugai/Roche, Tokyo, Japan) is a non-animated native-type produced by genetically engineered bacterial fermentation. Its spacing effect generally lasts for a few to several hours depending on its concentration and the anatomic factors of the injected site. Contrast medium (5% Iopamiron 300mgI/ml, Bayer, Germany) is commonly added to saline mixture for visualization on X-ray CT images. Because its durability is concentration-dependent, enriched gel is prepared on demand.

2.2 Assessment of indication and eligibility

To evaluate indications and the probability of effect and adverse effect, it is important to clarify previously irradiated doses and volumes in the involved at-risk organs, elapsed time after radiotherapy, and the present condition of the organs. In the case of no precise dose being available, a replicative simulation is required. According to published data, in most cases we may expect a certain degree of tissue recovery over the 6 months from the initial radiation [Abusaris, 2011; Nieder, 2006]. In addition to past history, the current status of at-risk organs is thought to be a significant determinant in assessing the vulnerability of the "recovered" status: hyperemia, edema, fibrosis, infection, hypoxia, atrophy, and other local pathophysiological conditions must be clarified and carefully interpreted. When any of these deteriorations is present, sufficient dose reduction is of greater importance.

2.3 Techniques of image-guided interventions and radiotherapy

2.3.1 Preparation of hyaluronate gel mixture

Premedication: Because the procedure is minimally invasive, sedative premedication is not mandatory, and in most cases an ordinary meal is recommended the morning of treatment to promote relaxation in the patient. Anxiolytic use of hydroxyzine pamoate or similar is recommended. However, careful individual medical management is required throughout the procedure in frail patients.

Keeping the patient awake: in terms of safety and precision, it is helpful for the patient to be able to report sensation during crucial needle maneuvers and during injection of the gel; therefore, deep sedation is not recommended if avoidable.

Use of sedatives: after completion of needle deployment, we recommend using benzodiazepam, which is effective in reducing post-traumatic psychological processing of the pain, anxiety, or fear immediately before and during the procedure.

Monitoring: the patient is connected to ECG, respiration, oxygen saturation, and blood pressure monitors, and covered with sterile drapes on the X-ray CT couch.

2.3.2 Needle deployment

2.3.2.1 Procedure

CT land marking and local anesthesia: puncture points are determined using the first set of plain CT images covering the target, and 10–20 ml of 1% lidocaine solution is then injected into the subcutaneous space at the intended puncture site. As well as superficial skin markings, tentative placement of a few 23 G needles inserted to a depth of 2–6 cm around the target as internal landmarks is useful in geometric planning to a deep target under X-ray CT guidance.

We use Microselectron system applicator needles: 1.1 mm in external diameter and 8–20 cm in length. The applicator needles are inserted following the landmark needles and deployed under X-ray CT or real-time ultrasound guidance. The needles are advanced into (or minimally over, if necessary) the tumor and deployed as indicated. Fine-pitch (2 or 3 mm thick) X-ray CT images are then acquired and transferred to the treatment planning computer (Plato, Nucletron, Veenendaal, Netherlands).

2.3.2.2 Special technical remarks regarding needle deployment

a. Bone and perivertebral intervention: Bone involvement is common and is a leading cause of cancer pain, either sensory or neuropathic. It is generally not difficult to pierce the involved bone through a window in which the healthy bony structure has been lost, irrespective of the existence of a calcified zone. A drill (e.g., a 13-gauge bone biopsy needle: Osteosite Bone Biopsy Needle; Cook, Canada) may be required to gain access through hard bone surface. In case of perivertebral needle insertion beneath the pleura, subpleural gel injection produces a safe space for needle insertion avoiding pulmonary.
b. Needle-tip-eye's view: To negotiate a step-by-step safe path to a deep target, we build a scenario of what the needle-tip encounters (e.g., skin, subcutaneous adipose tissue, muscle fascia, muscle, inter-fascial adipose tissue, and possibly a nerve plexus) before reaching the target.
c. Superficial lesions: We use a high-precision ultrasound system with 0.7 and 5.0 MHz (mostly 3.0-3.5 MHz) to examine superficial structures such as the dermis, epidermis, subcutaneous tissue, muscle, peritoneum, sub-fascial soft tissues, intestines, intestinal peristalsis, eyes, and vessels, as well as tumors; and to determine the infiltration range.
d. Nerve plexus: The patient may experience suddenly increased pain when pressure from the needle insertion cause further stimulation to a nerve plexus. A swollen tumor causing pressure may be covered with a layer of irritated nerves; in this case, additional local anesthesia with lidocaine is generally effective. Use of denervative anesthesia with

phenol glycerol may be chosen here, but with the effect of masking future signals of local tumor recurrence.

2.4 Hyaluronic gel injection

2.4.1 Preparation

The appropriate volume, concentration, and timing of injection of hyaluronic acid gel depends on the anatomical structure of the site and the required duration of spacing. Higher concentration provides spacing time. Most commonly, 50–150 mL of 1 mg/ml hyaluronic acid saline solution is prepared. A concentration of 1 mg/mL is sufficient for injection immediately before the start of irradiation, lasting in most cases from 30 minutes to 4 hours, depending on the site and structure of injection.

2.4.2 Injection

Injection into the space between the tumor and the risk organs is continued until a large enough thickness is obtained to create a margin of safety for the risk organ. If necessary, further injection is performed to maintain the space. Patency of the distance should be reconfirmed immediately before and after irradiation.

2.4.3 Confirmation

The space created by the injected gel is monitored by ultrasound or X-ray CT to check for unexpected gel migration and to measure the thickness of the gel space.

2.5 Brachytherapy

2.5.1 Trade-off principal

The individual prescribed dose is generally a trade-off between the target dose and the dose to risk organs to avoid serious complications,.

2.5.2 Treatment planning

After contouring the target and surrounding critical organs, and recognizing the needle positions on the X-ray CT images imported into the planning computer, a treatment plan is created using 3D-inverse planning with graphic optimization (Plato version 13.7, Nucletron).

2.5.3 Evaluation of dose–volume histogram

We assessed the desirable separation distances between the target and the risk organs, applying calculated total dose of the risk organs [Burman, 1991]. The target dose, risk organ dose, separation distance, and each patient's condition are individual trade-off factors. When it is difficult to prescribe the desirable dose, due to a low allowance for risk organ dose or due to difficulties encountered in creating a sufficient separation distance, the prescribed dose must be decreased and the distance recalculated step-by-step.

2.5.4 Estimation of equivalent dose

Equivalent dose in a conventional 2-Gy fraction schedule is calculated using the linear–quadratic (LQ) model and the linear–quadratic–linear (LQL) model [Astrahan, 2008], and is expressed as $GyE_{LQ2,\ \alpha/\beta=3}$ and $GyE_{LQL2,\ \alpha/\beta=3,}$ transition dose (DT) = 6, respectively, where $\alpha/\beta = 3$ and DT = 6 for late effects of tumor and normal tissue. It should be noted that the LQ model may not provide a precise estimation for single-shot irradiation [Brenner, 2008; Kirkpatrick, 2008]. In the setting ($\alpha/\beta = 3$), each single fraction dose of 20, 18, and 16 Gy was calculated to have an equivalent total dose of 92.0, 75.6, and 60.6 Gy, respectively, in a conventional 2-Gy fraction schedule.

2.6 Irradiation

After determining an optimal set of prescribed dose and separation distance parameters, the planning data are then sent to a remote after-loader system (Microselectron HDR Ir-192 version 2, Nucletron). In the present case series, a median dose of 18 Gy (range, 16–20 Gy) for GTV and a separation distance of not less than 1 cm was prescribed. The patient is moved to a shielded brachytherapy room and monitored by video-camera during the irradiation; vital data are also recorded. Irradiation generally takes 10–50 minutes, depending on the target size and dose.

2.7 Recovery to discharge

After the irradiation, we disconnect the patient from the monitoring equipment and remove the dwelling needles. Hemostasis is usually easy. The patient is moved to a couch to rest before being discharged.

2.8 Evaluation and follow-up

Patients are followed up at our outpatient clinic. Follow-up includes individual surveillance and management of adverse effects from the radiotherapy and the development of recurrent or metastatic disease. Tumor status and the patient's general condition are evaluated regularly. The first tumor evaluation is usually done 2 to 3 months after treatment. Any signs or symptoms of radiation-induced late toxicity are recorded and graded according to the latest common terminology criteria for adverse events [Trotti, 2003]. Pregabalin alone or in combination with opioids may be effective for residual or radiation-induced neuropathic pain [O'Connor, 2009].

3. Summary of clinical effects in the first 30 patients

We previously reported a series of 30 patients who were treated with reirradiation [Kishi, 2009]. The patients had received previous irradiation with a median dose of 60 Gy (range, 44–70 Gy) in 2-Gy fractions. All patients had subjective symptoms: 25 had pain, of which 21 was refractory to analgesics, and 25 complained of a local mass of which 3 were ulcerated. By location, there were 13 head and neck lesions, 3 breast, and 10 abdominal wall lesions: 2 each of bone, perineum, chest wall, intramuscular, and lymph nodal lesions; and 1 of the pelvic wall. Immediately before reirradiation, the median tumor (target) size was 4.0 cm (range, 2–13 cm), and median tumor volume was 18.8 cm^3 (range, 2.4–646.7 cm^3).

Of the 30 patients, 15 had locoregional recurrence of the primary disease and 10 had distant metastasis (of which 3 were incision disseminations and 5 were regional nodal relapse). By number of lesions at the time of reirradiation, 13 patients had one, 8 patients had two, 6 patients had three, and 3 patients had four. The at-risk organs involved 29 skin areas, 4 intestinal, 3 mucosal, and 3 retinal areas. The median dose to these at-risk organs in the previous radiotherapy was 50 Gy (range: 40–70 Gy) in 22–35 fractions. A single-fraction dose of 18.0 Gy (median, equivalent to 75.6 Gy at an α/β value of 3; range, 16–20 Gy) was prescribed to the tumor. The median created distance was 10 mm (range, 10–15 mm). Irradiation required approximately 10–50 minutes depending on the target size and dose. Gel injection resulted in a decrease in the dose to at-risk organs from 9.1 ± 0.9 Gy to 4.4 ± 0.4 Gy (mean ± standard deviation, $p < 0.01$), and the probability of normal tissue complications decreased from 60.8% ± 12.6% to 16.1% ± 19.8% ($p < 0.01$). We observed distinct tumor shrinkage in 20 of 21 eligible patients, including tumor disappearance in 6 patients; pain reduction in 18 of 21 eligible patients; and no unexpected late toxicity greater than grade 2.

The median observation period was 19.5 months (range, 3–43 months). At the second month, a distinct decrease in tumor volume was observed in 20 (95%) of 21 eligible patients, including 6 (24%) of 25 eligible patients in whom the tumor had disappeared.

4. Application to various body sites

4.1 Head and neck reirradiation

a. General skin and mucosal preservation

Background: Prudent skin and mucosa preservation is generally required in reirradiation treatment for recurrent head and neck cancer after initial radiotherapy. The involved skin and mucosa can be moved away from the target when subcutaneous or submucosal gel injection is available. We can generally expect an additional distance of 1 cm or more, which will decrease the skin/mucosal dose by approximately 50% of that without HGI, and the thickness can easily be increased by further injection.

Case and technique presentation: An 84-year-old male patient with hemiparalysis after brain infarction developed a submandibular lymphnode metastasis from oral floor cancer. We injected gel into the submucosal and subcutaneous space surrounding the tumor to create a safe distance, and then delivered 18Gy (76.5GyE$_{LQ2,\ \alpha/\beta=3}$) (Fig. 3).

b. Reirradiation to retropharyngeal lymphnode metastasis (mucosal preservation)

Background: Retropharyngeal node metastasis (RPNM) commonly occurs in the clinical course of head and neck cancers. The incidence of RPNM in oropharyngeal cancer has been reported as 27.5% [Gross, 2004], and post-surgical development of RPNM in hypopharyngeal cancer as 13.2% [Kamiyama, 2009]. This involvement has a significantly negative impact on the prognosis [Dirix, 2006], and there has been no safe and effective non-invasive or minimally invasive treatment for recurrence after radiotherapy.

Case and technique presentation: A left retropharyngeal lymph node (Rouviere's node) metastasis was found in a 50-year-old man who complained of an abnormal sensation in his neck 1 year after left neck surgery for pharyngeal pouch cancer followed by 50 Gy of external beam radiotherapy. Despite chemotherapy, the tumor increased in size and the symptoms worsened. The patient received reirradiation, as shown in Fig. 4, and has been healthy without any evidence of relapse for 4 years.

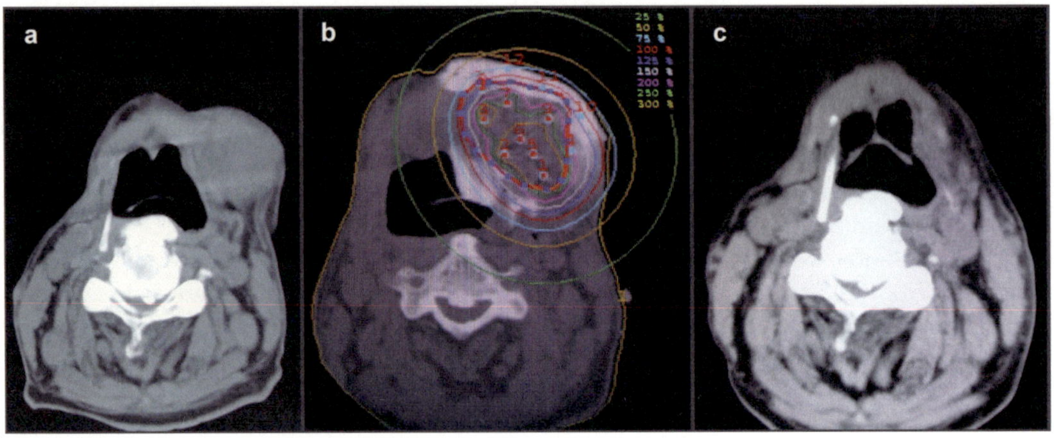

(a) Submandibular tumour before treatment. (b) The created distance measures approximately 1 cm. Injected gel is contrast enhanced. 18 Gy was delivered in a single fraction. (c) Six months after the brachytherapy.

Fig. 3. Skin and mucosal preservation in HDRBT by HGI.

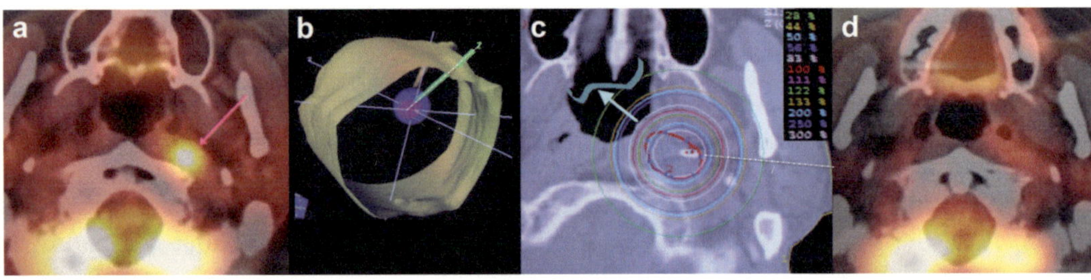

(a). An applicator needle (dotted line) was inserted from the left side of the neck (b) and a single dose of 20 Gy (100% line) was prescribed (c). Part of the needle can be seen in the target. Note the higher-dose area at the target. The pharyngeal mucosa was shifted to the low-dose area following injection of hyaluronate during irradiation (thick arrow and ribbon). There is a remarkable decrease in FDG accumulation 2 months after the brachytherapy (d). This photograph has been published previously [Kishi, 2009].

Fig. 4. Reirradiation to retropharyngeal lymphnode metastasis. PET study reveals abnormal accumulation of FDG in the metastatic retropharyngeal lymphnode

4.2 Abdominal/chest reirradiation

Numerous sites can be treated with this technique: chest wall, abdominal wall, various metastatic lymph nodes, adrenal gland metastasis, and parenchymal lesions. To date, curative reirradiation to paraaortic lymphnode metastasis has been one of the most difficult to safely achieve.

a. Reirradiation to paraaortic lymphnode metastasis:

Background: Paraaortic lymph nodal (PALN) metastasis is a frequent clinical sequel in various abdominopelvic malignancies, including pancreatic cancer, for which the rationale for surgery is limited [Pavlidis, 2011]. Eradicative PALN reirradiation would be ideal, but a safe dose is usually limited by surrounding radiosensitive organs, particularly the intestines. In reirradiation, the spinal cord and kidneys, as well as the intestines, may narrow the range of available beam angles. To date, only a few reports have described reirradiation in the abdomen, and these report the use of palliative doses [Chou, 2001; Haque, 2009].

Case and technique presentation: We encountered a patient with pancreatic cancer who developed PALN metastasis as an in-field recurrence, 6 months after resection of pancreatic cancer with 50 Gy of preoperative radiotherapy. A single fraction of 18 Gy was delivered to the tumor (75.6 Gy equivalent in a conventional schedule calculated with the LQ model at $\alpha/\beta = 2$ for the small intestines) and a total estimated D2cc (the minimum dose to the most irradiated volume of 2 cc) in the small intestines of 58.5 GyE $_{LQ2,\ \alpha/\beta = 3}$ with HGI; and 96 GyE $_{LQ2,\alpha/\beta = 3}$ without HGI. No complications were observed for 6 months. Three months after treatment, there was no FDG accumulation, tumor size had reduced, and the serum CA-19-9 value decreased from 5150 to 36.6 U/mL (normal range, <37.5). We consider that this case conclusively demonstrates that brachytherapy with HGI procedure by the paravertebral approach is safe and effective in reirradiation of PALN recurrence. The dosimetric and technical details of this case are to be published in the *Journal of Radiation Research* [Kishi, 2011b].

The paravertebral brachytherapy approach used in this case involved the application of a safe and reliable interventional procedure [Knelson, 1989] that is substantially unaffected by respiratory movement. This technique, under step-by-step X-ray CT guidance, achieved stable, precise needle deployment and gel injection.

4.3 Pelvic tumors

Also in the pelvic area, HDRBT with HGI may facilitate safe and effective reirradiation for recurrent primary lesions, lymph nodal or bone metastasis that have developed after radiotherapy to these areas [Kishi, 2011a]

a. Reirradiation to recurrent prostate cancer

In the treatment of prostate cancer, technological advancements in radiotherapy have enabled dose escalation to be performed safely. In intermediate- or high-risk prostate cancers, however, the reported local recurrence rate of 10%–30% remains high. A technique is required for further safe dose escalation for these refractory cancers, as well as a safe reirradiation technique for relapsed prostate cancer. Because the rectum is closely adjacent to the prostate, the radiotolerance limit of the rectum is generally lower than normal following curative external beam radiotherapy.

According to a summary in a review [van Vulpen, 2011] 5-year biological disease-free survival in six studies on reirradiation of prostate cancer was approximately only one third following initial curative dose treatment, and the incidences of grade 3 toxicities in the 11 studies on total 290 patients was approximately 4% of gastrointestinal toxicity and 14% of genitourinary toxicity, respectively. Further increase of toxicity should not be allowed in future attempts of dose escalation.

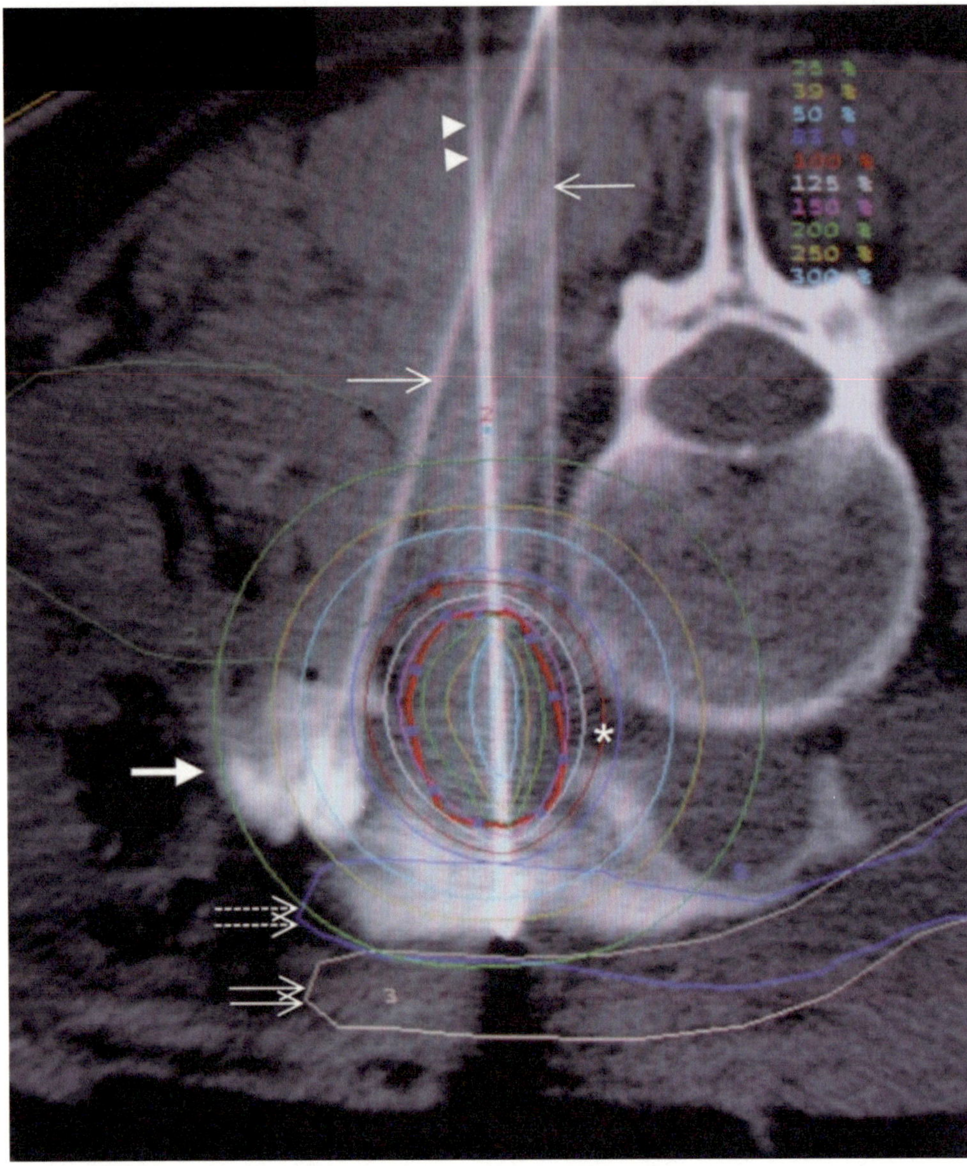

Fig. 5. Paravertebral approach and dose distribution. A small part of intestine was involved with more than 25% of the prescribed dose. Symbols: thick arrow, injected contrast-enhanced gel; thin arrows, injection needles; double arrowheads, brachytherapy applicator needles; double thin arrows, the closest part of small intestine shifted by HGI; double dotted arrows, superimposed lines of the original position of the closest part of small intestine; and thick line with dots, tumor contour. The oblique injection needle was inserted to move the second closest part of the small intestine, which was shifted aside. Isodose curves are 25%, 50%, 75%, 100% (asterisk), 125%, 150%, 200%, 250%, and 300%, in order from outermost to innermost. This figure taken from the original figures to be published in the *Journal of Radiation Research* [Kishi, 2011b].

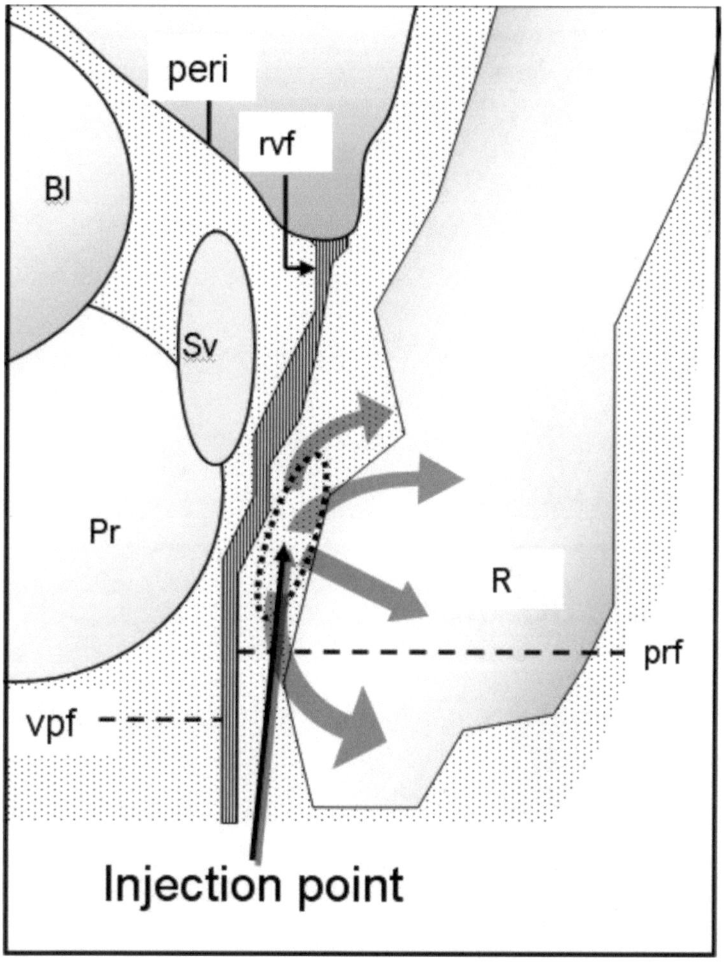

Fig. 6. Diagram showing the track of the injection needle to the injection point (dotted circle) in the anterior perirectal adipose tissue behind the rectal side of the rectovesical fascia (Denonvilliers' fascia). Gel injected from this point (half-tone arrows) flows into the perirectal and pararectal spaces to encase the rectum. Abbreviations (in alphabetical order): Bl, bladder; peri, peritoneum; Pr, prostate; prf, perirectal fascia; R, rectum; rvf, rectovesical fascia (Denonvilliers' fascia); Sv, seminal vesicle; vpf, vesicoprostatic fascia.

Case and technique presentation: In a patient with local recurrence of prostate cancer at 18 months after initial radiotherapy of 61.8 $GyE_{LQ2,\alpha/\beta=3}$ to the prostate, we prescribed 16 Gy (60.8 $GyE_{LQ2,\alpha/\beta=3}$ or 78.2 $GyE_{LQ2,\alpha/\beta=1.6}$) of reirradiation HDRBT with HGI (Figs. 6, 7). The procedure was achieved in 10 minutes, without complications. Rectal D2cc for reirradiation was 5.6 Gy (9.58 $GyE_{LQ2,\alpha/\beta=3}$). Compared with the initial radiation, the gel injection resulted in an improved therapeutic ratio. The patient was regularly followed up at our clinic: at over 3.5 years after reirradiation there was no evidence of recurrence or radiation-related toxicities greater than Grade 2, maintaining a nadir PSA level of 0.03 ng/ml without hormonal therapy. The details and treatment technique in this case have been reported in *Brachytherapy* [Kishi, 2011a].

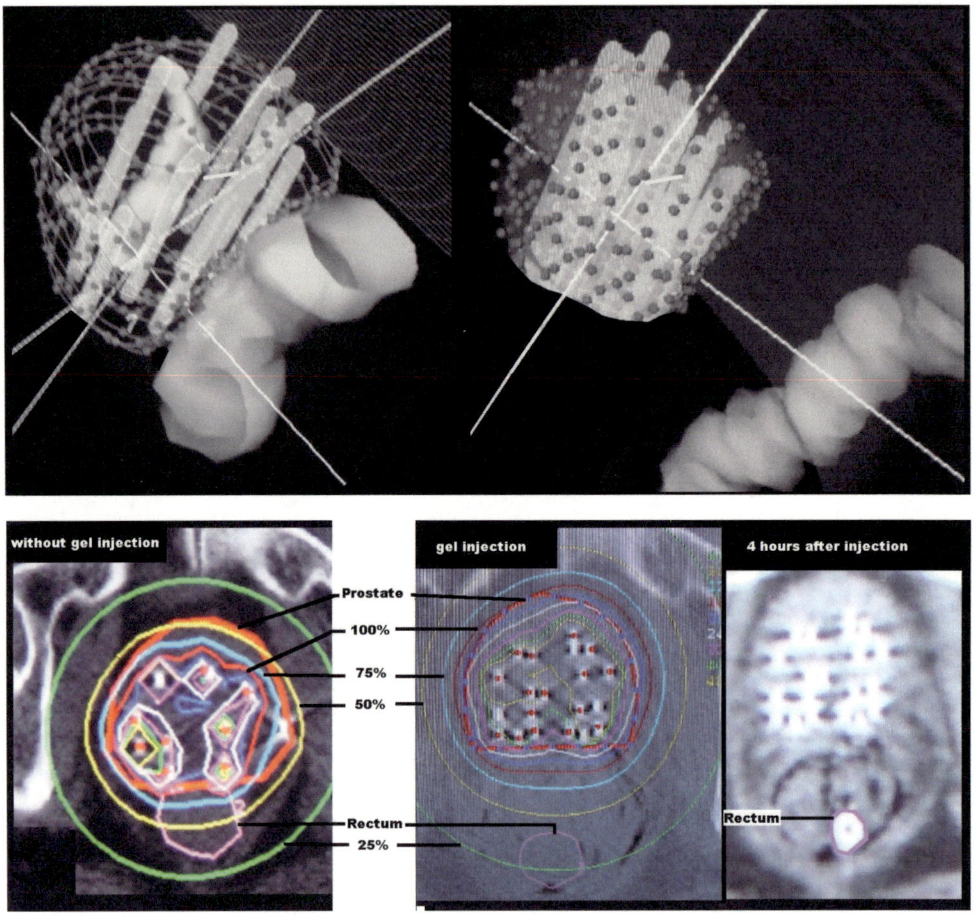

Fig. 7. Upper row: Reconstructed images of the rectum and the prostate for the initial (left) and reirradiation (right) brachytherapy (interval between dots on the prostate is 8 mm). Following gel injection, the rectum is shifted posteriorly and is reduced in size.

Lower row: Dose distribution curves for the initial (left) and reirradiation (middle) brachytherapies. The rectum and prostate are separated more than 25 mm (measureable with the dot interval). Thin layers of air are seen in and along the perirectal fascia. The relative positions of the rectum and prostate are maintained even 4 hours after injection (right). Contrast medium was injected into the rectum. The figures were form the original manuscript for that published in Brachytherapy [Kishi, 2011a].

5. Discussion

5.1 Problems to be addressed in reirradiation treatment

5.1.1 Hypoxic environment of reirradiation target

Strong hypoxia in recurrent tumor after irradiation: Tumors larger than a certain size (>0.1 mm) usually develop hypoxic areas due to insufficient vascular supply. These areas are also a sanctuary from hydrophilic chemotherapeutic agents, although launching waves of attacks may be effective. Cellular adaption to hypoxia includes cell cycle arrest and resulting prolongation of repair deadlines, increased capability of anaerobic glycolysis resulting in lactic acid pooling, and cellular immigration programs towards more oxygen rich environments producing actins and losing cadherins. Previously irradiated tissue tends to be more hypoxic and therefore frequently more radioresistant. Popovtzer et al. reported 92% of the third recurrence was the in-field of the second irradiation (reirradiation), which occurred in 77% of median 68-Gy-reirradiated patients in 154 relapsed squamous head and neck cancer patients [Popovtzer, 2009]. This suggests much higher dose may be biologically required to overcome the reparability of recurrent tumor after initial radiotherapy. In this situation, HDRBT has the important advantage of providing high-dose areas close to the source inserted in the tumor. Furthermore, another possibility that an incomplete reirradiation may promote the immigration capability should be taken into account.

5.1.2 Diagnostic jeopardy

Precise judgment of a diagnosis of recurrence can be difficult because of tissue hypoxia induced by the radiotherapy itself and the biological responses that occur in and around the irradiated area. Diagnostic features for tumor recurrence includes, 1) increase in tumor size (especially new nodule protrusion), 2) revascularization in the tumor (not static revascularization around the tumor), and 3) strong increase in FDG-PET accumulation. Swelling of the tumor, emergence of a ring enhancement zone, and persistent weak or moderate appearance of FDG-PET accumulation in or around the irradiated tumor are often observed in reactive inflammatory phenomena, or tumor-related. Currently recommendable approach to make a differential diagnosis in these situations is only to rationally integrate and interpret the time course of PET-CT, enhanced MR, and MR diffusion imaging findings as well as the biological nature of the tumor cells as well as post-irradiation changes of involved normal tissues.

5.1.3 Occurrence of complications

In our previous report of reirradiation with a total accumulated target dose of 125.6 Gy (median)[Kishi, 2009], the actual incidence of complications was very low (0% directly related to the radiotherapy) than expected (16.1%) over a 19.5-month observation period. This may have occurred because we had created a sufficient safety margin in the gel spacing procedure, and because of the time interval from the previous radiation effect.

5.1.4 The problem of equivalent dose calculation

The LQ model may overestimate a single high-dose effect: the LQ model is well validated, experimentally and theoretically, allegedly up to approximately 10 Gy/fraction, and is reasonable for clinical use up to a large dose of about 18 Gy per fraction in fractionated schedules [Brenner, 2008]. Recent studies revealed the significant overestimation of large single-dose treatment effect calculated by the LQ model, in both clinical and experimental data [Astrahan, 2008; Iwata, 1984; Kirkpatrick, 2008]. Overestimation of the single high-dose effect tends to be greater in a dose larger than DT = 2 x (α/β ratio) [Astrahan, 2008]. In the LQL model, a linear function works over the range of DT Gy. The LQL model is currently under evaluation.

5.1.5 Interval repair

An interval of more than 6 months was considered to increase spinal tolerance level [Nieder, 2005; 2006]. Several authors reported the effectiveness of reirradiation by external beam [Okamoto, 2002; Oksuz, 2004; Sulman, 2008; Wu, 2007; Wurschmidt, 2008] with a relatively small incidence of grade 3–4 toxicity for a cumulative dose of 110–117 Gy, where the median intervals ranged from 13 to 92 months. However we must be careful to apply some idea of interval repair because an individual tissue tolerance differs not only by the dose and the interval.

5.2 Injectable material

1. Artificial cross-linked hyaluronate: There are long-lasting variants of hyaluronate that are artificially cross-linked and resistant to biodegration, some of which are durable for months (Restylane SubQ, Q-med, Uppsala, SWEDEN)[Restylane®-International] and are used as a filler by subcutaneous injection in cosmetic augmentations. Unlike native-type hyaluronate, no biological research, including the receptor-mediated process, is available for this variant. Use of this type of hyaluronate was reported by Prada et al. for creating and maintaining a space during IMRT, HDRBT, and LDRBT for prostate cancer [Prada, 2007, 2009]. In response, Vordermark et al. proposed that a material with a faster resolution would be suitable for application to HDRIBT [Vordermark, 2008]. These long-lasting hyaluronate implants may cause immune [Hamilton, 2007] or inflammatory reactions [Arron, 2007], or infections [Christensen, 2009] [Arron, 2007; Edwards, 2007; Ghislanzoni, 2006; Wiest, 2009; Wolfram, 2006], and have been surgically removed from some patients [Arron, 2007; Edwards, 2007; Ghislanzoni, 2006; Wiest, 2009; Wolfram, 2006].
2. Dextrose solution: Dextrose solution is one of the most commonly used spacers. The low viscosity of the fluid makes it difficult to create an effective durable space, meaning that a large dose is generally required. A previous study reported that 625 ml (range, 250–1,200 mL) of 5% dextrose in water was required to separate organs (by approximately 1 cm according to the published figures) during radiofrequency ablation [Arellano, 2010]. Electrolyte imbalance and fluid overload must be carefully monitored in this situation.

5.3 Time and cost-effectiveness of the present method

The cost of native-type high-molecular-weight hyaluronate is US$10–US$15 for a 2.5 ml (2.5 mg) vial, which is approximately one-sixtieth the cost of the artificially cross-linked durable type. The time required to complete the procedure is 10–15 minutes in most cases; thus, HGI is highly time- and cost-effective.

5.4 Further possibilities

As well as its use for creating a safe distance, or enabling curative radiotherapy that has been impossible until now, HGI has potential for further development. This spacing method will safely guide the insertion of larger therapeutic devices and/or organs without requiring an intensive surgical procedure. This technique has the potential to affect a profound shift in the manner in which physicians consider retreatment options for previously irradiated tissue, in a manner hitherto unknown.

6. Summary and conclusion

6.1 Background

In this chapter we describe a technical solution for problems associated with reirradiation. Radiotherapy is singularly effective in cancer control, even in a second treatment (reirradiation). Sufficient dose delivery may promise a local cure, but reirradiation is usually limited due to the tolerance level of surrounding normal tissue.

6.2 Method

To safely perform high-dose reirradiation treatment, our radiation oncologists and interventional radiologists cooperated to develop a novel procedure based on a recent advance in molecular science. Our aim was to create a safe distance in critical radiotherapy by injecting native-type high-molecular-weight hyaluronic acid between the target and organs at risk under ultrasound or X-ray CT guidance. The procedure is termed hyaluronic gel injection (HGI). The material protects tissues from injury and inflammation, and is reported in recent studies to inhibit cell migration and proliferation mediated by surface receptors including CD44. The injection is a quick and minimally invasive procedure performed with a 21-gauge needle, which, unlike surgery, enables a stepwise or fractionated schedule of outpatient treatment for multiple lesions.

During the period that the created distance is maintained by the injected gel, typically a few hours, a single-session irradiation is performed by high-dose-rate brachytherapy (HDRBT) with CT-based 3D planning. We reported that this injection procedure provides a significant reduction in normal tissue complication probability (NTCP) in various situations, enabling doses ranging from 15 to 20 Gy to be safely delivered to the target without significantly involving surrounding at-risk organs.

6.3 Result

The biological equivalent dose for the dose range was 50.4–92 Gy for a conventional radiotherapy schedule at $\alpha/\beta = 3$, fulfilling the individual dose requisition for each curative purpose. In comparative analysis of our clinical records, we found that the therapeutic ratio (TR) of target dose to risk-organ dose was increased by approximately three times by this HGI procedure. To the best of our knowledge, other than that for surgery, this enhancement factor of TR is the largest value found to date. In our published report of 30 patients with recurrent cancer after 60 Gy (median) of previous external beam treatment, reirradiation by HDRBT with HGI resulted in distinct tumor shrinkage in 95%; and significant pain reduction, and Grade 2 or larger early and late toxicity in 95%, 85%, and 0%, respectively, during the 19.5-month observation period. To date, we have reported reirradiation in cases of paraaortic lymphnode metastasis, recurrent lung cancer, uterine cancer, prostate cancer, and head and neck cancer, among others, with long-lasting curative effect.

6.4 Conclusion

This risk-organ-sparing preservation procedure offers safe and efficient reirradiation treatment for recurrent cancer patients, in terms of longstanding local cancer control without significant physical stress, as well as providing pain reduction.

7. Acknowledgements

This research and publication was supported in part by a Grant-in-Aid for Scientific Research from the Ministry of Education, Culture, Sports, Science and Technology, Japan (MEXT Grant), grant number 23659595.

8. References

Abusaris, H., P.R. Storchi, R.P. Brandwijk, &J.J. Nuyttens, (2011(epub)). Second re-irradiation: Efficacy, dose and toxicity in patients who received three courses of radiotherapy with overlapping fields. *Radiother Oncol,* Vol. 99,No.2, pp. 235-239

Arellano, R.S., V.L. Flanders, S.I. Lee, P.R. Mueller, &D.A. Gervais, (2010). Imaging-guided percutaneous radiofrequency ablation of retroperitoneal metastatic disease in patients with gynecologic malignancies: clinical experience with eight patients. *AJR Am J Roentgenol,* Vol. 194,No.6, pp. 1635-1638

Arron, S.T. and I.M. Neuhaus, (2007). Persistent delayed-type hypersensitivity reaction to injectable non-animal-stabilized hyaluronic acid. *J Cosmet Dermatol,* Vol. 6,No.3, pp. 167-171

Astrahan, M. (2008). Some implications of linear-quadratic-linear radiation dose-response with regard to hypofractionation. *Med Phys,* Vol. 35,No.9, pp. 4161-4172

Brenner, D.J. (2008). The linear-quadratic model is an appropriate methodology for determining isoeffective doses at large doses per fraction. *Semin Radiat Oncol,* Vol. 18,No.4, pp. 234-239

Burman, C., G.J. Kutcher, B. Emami, &M. Goitein, (1991). Fitting of normal tissue tolerance data to an analytic function. *Int J Radiat Oncol Biol Phys,* Vol. 21,No.1, pp. 123-135

Campo, G., A. Avenoso, S. Campo, A. D'Ascola, P. Traina, D. Samǽ□, &A. Calatroni, (2008). The antioxidant effect exerted by TGF-1beta-stimulated hyaluronan production reduced NF-kB activation and apoptosis in human fibroblasts exposed to FeSo4 plus ascorbate. *Mol Cell Biochem,* Vol. 311,No.1-2, pp. 167-177

Chou, H.H., C.C. Wang, C.H. Lai, J.H. Hong, K.K. Ng, T.C. Chang, C.J. Tseng, C.S. Tsai, &J.T. Chang, (2001). Isolated paraaortic lymph node recurrence after definitive irradiation for cervical carcinoma. *Int J Radiat Oncol Biol Phys,* Vol. 51,No.2, pp. 442-448

Christensen, L.H., (2009). Host tissue interaction, fate, and risks of degradable and nondegradable gel fillers. *Dermatol Surg,* Vol. 35 Suppl 21612-1619

Damast, S., J. Wright, M. Bilsky, M. Hsu, Z. Zhang, M. Lovelock, B. Cox, J. Zatcky, &Y. Yamada, (2010). Impact of Dose on Local Failure Rates after Image-Guided Reirradiation of Recurrent Paraspinal Metastases. *Int J Radiat Oncol Biol Phys,* Vol.

Dirix, P., S. Nuyts, B. Bussels, R. Hermans, &W. Van den Bogaert, (2006). Prognostic influence of retropharyngeal lymph node metastasis in squamous cell carcinoma of the oropharynx. *International Journal of Radiation Oncology*Biology*Physics,* Vol. 65,No.3, pp. 739-744

Edwards, P.C. and J.E. Fantasia, (2007). Review of long-term adverse effects associated with the use of chemically-modified animal and nonanimal source hyaluronic acid dermal fillers. *Clin Interv Aging,* Vol. 2,No.4, pp. 509-519

Emami, B., J. Lyman, A. Brown, L. Coia, M. Goitein, J.E. Munzenrider, B. Shank, L.J. Solin, &M. Wesson, (1991). Tolerance of normal tissue to therapeutic irradiation. *Int J Radiat Oncol Biol Phys,* Vol. 21,No.1, pp. 109-122

Ghislanzoni, M., F. Bianchi, M. Barbareschi, &E. Alessi, (2006). Cutaneous granulomatous reaction to injectable hyaluronic acid gel. *Br J Dermatol,* Vol. 154,No.4, pp. 755-758

Gross, N.D., T.W. Ellingson, M.K. Wax, J.I. Cohen, &P.E. Andersen, (2004). Impact of retropharyngeal lymph node metastasis in head and neck squamous cell carcinoma. *Arch Otolaryngol Head Neck Surg,* Vol. 130,No.2, pp. 169-173

Hamilton, R., J. Strobos, &N. Adkinson, (2007). Immunogenicity studies of cosmetically administered nonanimal-stabilized hyaluronic acid particles. *Dermatol Surg,* Vol. 33,No.suppl.2, pp. s176-s185

Haque, W., C.H. Crane, S. Krishnan, M.E. Delclos, M. Javle, C.R. Garrett, R.A. Wolff, &P. Das, (2009). Reirradiation to the abdomen for gastrointestinal malignancies. *Radiat Oncol,* Vol. 455

Hashizume, M. and M. Mihara, (2010). High molecular weight hyaluronic acid inhibits IL-6-induced MMP production from human chondrocytes by up-regulating the ERK inhibitor, MKP-1. *Biochem Biophys Res Commun,* Vol. 403,No.2, pp. 184-189

Hayashi, S., H. Hoshi, &T. Iida, (2002). Reirradiation with local-field radiotherapy for painful bone metastases. *Radiat Med,* Vol. 20,No.5, pp. 231-236

Iwata, S., S. Miyauchi, &M. Takehana, (1984). Biochemical studies on the use of sodium hyaluronate in the anterior eye segment. I. Variation of protein and ascorbic acid concentration in rabbit aqueous humor. *Curr Eye Res,* Vol. 3,No.4, pp. 605-610

Joseph, K.J., Z. Al-Mandhari, N. Pervez, M. Parliament, J. Wu, S. Ghosh, P. Tai, J. Lian, &W. Levin, (2008). Reirradiation After Radical Radiation Therapy: A Survey of Patterns of Practice Among Canadian Radiation Oncologists. *Int J Radiat Oncol Biol Phys,* Vol.

Kamiyama, R., M. Saikawa, &S. Kishimoto, (2009). Significance of retropharyngeal lymph node dissection in hypopharyngeal cancer. *Jpn J Clin Oncol,* Vol. 39,No.10, pp. 632-637

Kirkpatrick, J.P., J.J. Meyer, &L.B. Marks, (2008). The linear-quadratic model is inappropriate to model high dose per fraction effects in radiosurgery. *Semin Radiat Oncol,* Vol. 18,No.4, pp. 240-243

Kishi, K., H. Adati, &K. Takada, Implants for radiation therapy, in *Patent publication bulletin,* J.P. Office, Editor. 2004, Kuraray Medical: Japan.

Kishi, K., M. Sato, S. Shirai, T. Sonomura, &R. Yamama, (2011a). Reirradiation of prostate cancer with rectum preservation: Eradicative high-dose-rate brachytherapy with natural type hyaluronate injection. *Brachytherapy,* Vol.

Kishi, K., S. Shirai, M. Sato, &T. Sonomura, (2007a). Computer-aided preservation of risk organs in critical brachytherapy by tissue spacing with percutaneous injection of hyaluronic aid solution. *Int J Radiat Oncol Biol Phys,* Vol. 69,No.3, pp. S568-S569

Kishi, K., S. Shirai, M. Sato, T. Sonomura, &K. Tanaka. (2007b) Preservation of risk organs in critical brachytherapy by tissue spacing with percutaneous injection. in *The 5th Japan-US Cancer Therapy Symposium & The 5th S. Takahashi Memorial Joint Symposium.*: The 5th Japan-US Cancer Therapy Symposium & The 5th S. Takahashi Memorial Joint Symposium Committee. Sendai, Japan

Kishi, K., T. Sonomura, S. Shirai, Y. Noda, M. Sato, M. Kawai, &H. Yamaue, (2011b). Brachytherapy reirradiation with hyaluronate gel injection of paraaortic lymphnode metastasis of pancreatic cancer: paravertebral approach - A technical report with a case -. *J Radiat Res,* Vol. In press.

Kishi, K., T. Sonomura, S. Shirai, M. Sato, &K. Tanaka, (2009). Critical organ preservation in reirradiation brachytherapy by injectable spacer. *Int J Radiat Oncol Biol Phys,* Vol. 75,No.2, pp. 587-594

Kishi, K., K. Takifuji, S. Shirai, T. Sonomura, M. Sato, &H. Yamaue, (2006). Brachytherapy technique for abdominal wall metastases of colorectal cancer: ultrasound-guided insertion of applicator needle and a skin preservation method. *Acta Radiol,* Vol. 47,No.2, pp. 157-161

Knelson, M., J. Haaga, H. Lazarus, C. Ghosh, F. Abdul-Karim, &K. Sorenson, (1989). Computed tomography-guided retroperitoneal biopsies. *J Clin Oncol,* Vol. 7,No.8, pp. 1169-1173

Krasinski, R., H. Tchorzewski, &P. Lewkowicz, (2009). Antioxidant effect of hyaluronan on polymorphonuclear leukocyte-derived reactive oxygen species is dependent on its molecular weight and concentration and mainly involves the extracellular space. *Postepy Hig Med Dosw (Online),* Vol. 63205-212

Kusunoki, M., H. Ikeuchi, H. Yanagi, M. Noda, H. Tonouchi, Y. Mohri, K. Uchida, Y. Inoue, M. Kobayashi, C. Miki, &T. Yamamura, (2005). Bioresorbable hyaluronate-carboxymethylcellulose membrane (Seprafilm) in surgery for rectal carcinoma: a prospective randomized clinical trial. *Surg Today,* Vol. 35,No.11, pp. 940-945

Nieder, C., A.L. Grosu, N.H. Andratschke, &M. Molls, (2005). Proposal of human spinal cord reirradiation dose based on collection of data from 40 patients. *Int J Radiat Oncol Biol Phys,* Vol. 61,No.3, pp. 851-855

Nieder, C., A.L. Grosu, N.H. Andratschke, &M. Molls, (2006). Update of human spinal cord reirradiation tolerance based on additional data from 38 patients. *Int J Radiat Oncol Biol Phys,* Vol. 66,No.5, pp. 1446-1449

O'Connor, A.B. and R.H. Dworkin, (2009). Treatment of neuropathic pain: an overview of recent guidelines. *Am J Med,* Vol. 122,No.10 Suppl, pp. S22-32

Okamoto, Y., M. Murakami, E. Yoden, R. Sasaki, Y. Okuno, T. Nakajima, &Y. Kuroda, (2002). Reirradiation for locally recurrent lung cancer previously treated with radiation therapy. *Int J Radiat Oncol Biol Phys,* Vol. 52,No.2, pp. 390-396

Oksuz, D.C., G. Meral, O. Uzel, P. Cagatay, &S. Turkan, (2004). Reirradiation for locally recurrent nasopharyngeal carcinoma: treatment results and prognostic factors. *Int J Radiat Oncol Biol Phys,* Vol. 60,No.2, pp. 388-394

Pavlidis, T.E., E.T. Pavlidis, &A.K. Sakantamis, Current opinion on lymphadenectomy in pancreatic cancer surgery. *Hepatobiliary Pancreat Dis Int,* Vol. 10,No.1, pp. 21-25

Popovtzer, A., I. Gluck, D.B. Chepeha, T.N. Teknos, J.S. Moyer, M.E. Prince, C.R. Bradford, &A. Eisbruch, (2009). The Pattern of Failure After Reirradiation of Recurrent Squamous Cell Head and Neck Cancer: Implications for Defining the Targets. *Int J Radiat Oncol Biol Phys,* Vol.

Prada, P.J., J. Fernandez, A.A. Martinez, A. de la Rua, J.M. Gonzalez, J.M. Fernandez, &G. Juan, (2007). Transperineal injection of hyaluronic acid in anterior perirectal fat to decrease rectal toxicity from radiation delivered with intensity modulated